Oil Supply Crisis

Oil Supply Crisis

Cooperation and Discord in the West

Vessela Chakarova

LEXINGTON BOOKS
Lanham • Boulder • New York • Toronto • Plymouth, UK

Published by Lexington Books
A wholly owned subsidiary of The Rowman & Littlefield Publishing Group, Inc.
4501 Forbes Boulevard, Suite 200, Lanham, Maryland 20706
www.rowman.com

10 Thornbury Road, Plymouth PL6 7PP, United Kingdom

British Library Cataloguing in Publication Information Available

Library of Congress Cataloging-in-Publication Data

Library of Congress Cataloging-in-Publication Data Available

ISBN 978-0-7391-6975-9 (cloth)—978-0-7391-6976-6 (electronic)
ISBN 978-1-4985-1543-6 (pbk)

Contents

List of Figures

List of Tables

Abbreviations

AIOC	Anglo-Iranian Oil Company
BP	British Petroleum
DoE	Department of Energy
EC	European Community
EPSC	European Petroleum Supply Committee
FPSC	Foreign Petroleum Supply Committee
IEA	International Energy Agency
IEP	International Energy Program
IIAB	International Industry Advisory Board
IPC	Iraq Petroleum Company
LNG	liquefied natural gas
mbd	million barrels per day
MEEC	Middle East Emergency Committee
MSs	Member States
NATO	North-Atlantic Treaty Organization
NIOC	National Iranian Oil Company
OECD	Organization for European Cooperation and Development
OELAC	Oil Emergency London Advisory Committee
OPEC	Organization of the Petroleum Exporting Countries
OPEG	OEEC Petroleum Emergency Group
OSAC	Oil Supply Advisory Committee

Preface

This study grew out of my doctoral research at Old Dominion University. I have always been interested in the work of international organizations—these relatively new actors on the international scene, but which significance has grown tremendously in the last sixty years. My master thesis focused on the involvement and attempts at mediation and conflict resolution of several IOs in the conflict provoked by the dissolution of Yugoslavia in the 1990s. Later on, I wrote about NATO and its strategy, and I attempted to theorize about European integration. The idea to write about cooperation during oil supply shortages came to me while I was reading about past oil crises and noticed that the level of cooperation differed significantly from one instance to another. Naturally, I asked the question under what conditions did the states affected by the shortage cooperate, in other words, why did it happen in some cases and not in others. This was, of course, puzzling, and I decided to investigate it further.

Let me briefly explain what this book is not about. While it provides quite a detailed account of the workings of the oil market, and considers different market variables that might have contributed to cooperative behavior, this study is not about the oil market. The goal of this research was to provide a test for two competing hypotheses of international cooperation, advanced by two major schools of international relations—neorealism and neoliberalism. As such, it is a theoretical work within the field of international relations, rather than an investigation of the oil market.

This volume would not have been possible without the support I received from colleagues, family and friends. My deepest gratitude is to my doctoral advisor Dr. Kurt Taylor Gaubatz who has been an amazing mentor guiding me through this arduous journey. He gave me invaluable support and advice, helping me to reach high standards of research and teaching. I have learned a lot from him. Drs. Steve Yetiv and Larry Filer were encouraging along the way and provided critical improvements to the final document. I want to thank the entire faculty of the International Studies Program at Old Dominion University, who have taught me a lot and have helped me become a scholar. I am also grateful to Nick Volnick, who took the time to read and correct my writing. My university colleagues and friends Matt Hall, Eva Svobodova, Stephanie Smith, Anna Rulska and Denita Acker were helpful with advice, sympathy and understanding. My husband, Arturo Tejada Ruiz, who has always been very supportive and encouraging, and who patiently endured my late-night work, my disappointments, and my lectures on energy crises and European politics. My sister Lidia Abrasheva, and my friends Bobbie Fuller, Marija Raskovic, Tim Chakraborty, Paul Schultze, Jose Padilla and Alexandra Parra helped by being there for me. And finally, my mother Valkana Chakarova and my father Plamen Chakarov, whom I lost along the way, but who has helped me become the person I am. I couldn't have done this without their guidance, and support.

Chapter 1
Introduction

Oil: A Critical Commodity

"Oil in the modern world might rightly be termed the life-blood of industry, agriculture and transport."[1] These are the starting words of a report on oil supply and consumption in Europe, published in 1956. They could well have been written, however, a year or a month ago. For our world has changed significantly in some respects and yet not much in others. Oil is still the lifeblood of modern economies. It is the one ingredient of modern life without which all society would come to a halt. It is the most covered and reported on of all commodities. Oil prices fluctuations are followed by a myriad of traders, economic and policy analysts, and market observers. They are an important element of consumer countries' security of supply, that is, the uninterrupted flow of oil at affordable prices. Oil supply security is especially critical for European countries, most of which have very few oil resources of their own. The EU currently imports more than 80% of its oil—a figure expected to increase in the future. This has serious implications for European economic growth, and for its relations with its neighbors, and the rest of the world. Europeans are frequently reminded of their energy dependency, notably by their most important energy partner—Russia. Russia-EU disputes have mostly involved natural gas, but some of them have spilled over oil, the latest one in January 2010, when Moscow reduced oil flows to Byelorussia, threatening supplies flowing through the Druzhba pipeline to Germany, Poland and three other countries. The crisis was quickly diffused but its bitter aftertaste reminds Europeans every day of their high vulnerability in the energy area.

And European countries are not the only ones concerned. Despite all our efforts, oil is still one of the most important fossil fuels, used for transportation, a variety of industrial processes and electricity production, among others. Renewable resources, such as wind, solar and biomass are far from satisfying even a fraction of our energy needs, and the recent backlash against nuclear power, after the Fukushima accident in Japan of 2011, is setting us further away from a fossil-free world. Thus, oil is here to stay for the foreseeable future. It is a relatively cheap resource, relatively easy to transport, and it is still the main fuel used for transportation, without which most economic and industrial activity would stop. All these characteristics of oil make it a very valuable resource, critical for a state's functioning and survival. This makes inter-state cooperation in the management of this resource very important. Ironically, this is also the reason why cooperative behavior is difficult to come by in this area.

This volume looks broadly at past oil supply crises and consumer countries' reaction to them. I examine more specifically the question of whether European governments coordinated their response to the oil shortage, and under what conditions coordination occurred. International relations theories offer two contending explanations of cooperation: the hegemonic stability thesis and institutionalism. The latter has attained a high degree of acceptance among scholars, as international organizations have proliferated creating high expectations for the future of inter-state cooperation. This claim is put to the test by comparing oil crisis cooperation within four international institutions. The study has a twofold purpose. First, it will provide an insight into the nature of cooperative behavior, and more specifically inter-state cooperation within international organizations. This is accomplished through contrasting the assumptions on international cooperation made by two mainstream international relations theories—neorealism and neoliberalism—thus, making a valuable contribution to one of the longest-standing debates in IR. Second, the study attempts to improve our understanding of the dynamics of oil disruptions, by studying in detail seven supply shortages. A specific emphasis will be placed on consumer reaction to the shortage. By outlining the nature of the consumer response and the particular conditions under which it evolved, the analysis will shed light on an important but often neglected aspect of oil politics—supply disruptions. For while the international system has evolved considerably in the last sixty years, oil supply disruptions are not a phenomenon of the past.

Oil Crises: Is There a Risk?

There are, currently, two diametrically opposed opinions on the matter. On one hand, there are the optimists who announce the end of our energy struggles, as we gradually embrace renewable energies, and reduce our use of oil. In addition, with the advance of technology, new oil sources, hitherto unreachable, become available. The oil market, the optimists argue, has evolved considerably since the 1970s, gradually becoming more transparent. Important innovations have been introduced along the way: deregulation of prices, spot rather than term contracts, futures trades, and an increase in government-controlled inventories. The Middle East, which holds a dominant share of global oil resources, has undergone a sort of transformation, as well. It is now more stable and less prone to disturbances of the 1973 type. The second group—the alarmists—keeps ringing the bell of peak oil. Oil is a finite resource, they say, which is diminishing. It is concentrated in the hands of a few corrupt regimes, making it very vulnerable to supply disruptions. In addition, the rise of China and India as major players on the oil market increases the competition for resources.

The latest evidence seems to be supporting the second group. Oil reserves will be increasingly concentrated in only one region—the Middle East—which political environment is very volatile. Non-OPEC supply is expected to decrease gradually starting from 2012, and most future production growth will be concentrated in OPEC countries.[2] Current production expansion projects will not match predicted growth in demand. In any case, most new oil will originate either from non-conventional sources or from deep off-shore fields, which are more expensive to explore and have higher CO2 emissions.

In addition, almost all spare capacity—an important hedge on the oil market—is currently concentrated in one country—Saudi Arabia. In 2010, the Saudis held on the average 3.76 mbd of surplus oil (4.3 % of world consumption) out of 4.36 mbd world surplus capacity.[3] The optimists argue that Saudi Arabia is a reliable ally of the West and is part of the "doves" wing of OPEC. The Saudi ruling family—realizing that keeping consumers well supplied is important for producers' long-term well-being, has indeed always taken a pragmatic view regarding world oil. For example, throughout the 1970s, Saudi oil minister Yamani repeatedly warned fellow producers that depriving Westerners of oil is a two-edged sword, as it will lead to more conservation and fuel-switching, and thereby serve to reduce their dependence on oil. The Saudi psychology has probably not changed, but with the rise of organized transnational terrorism, the Middle East has. Let us recall that fifteen of the nineteen September 11 hijackers were from Saudi Arabia. There is also evidence that a variety of Saudi charities and Muslim foundations have financially supported Al Qaeda.[4] There is still a lot of opposition in the country to the royal family's warm relations with the West. A number of terrorist attacks have happened on Saudi territory, as well. And while the government has been firmly committed to fighting terrorist networks on its territory, there is no guarantee that the next terrorist attack will spare Saudi oil installations. In addition, the specter of a Saudi uprising, similar to the Arab spring movements of 2011, is always in the air. Aware of this possibility, the Saudi king Abdullah recently introduced a number of reforms, including anti-corruption measures and voting rights for women. It still remains to be seen, however, whether these superficial changes will satisfy a predominantly young and disillusioned population.

A reduction of Saudi oil supplies will impact the market tremendously. Saudi Arabia supplied 12.8% of global oil (crude and products) in 2011. Not only is the country the biggest oil producer, it also has the highest world oil reserves, estimated currently at 266 billion barrels.[5] A major interruption of Saudi oil supplies would create chaos on the oil market of unimaginable proportions.

The Saudis are not the only potential source of turmoil in the Middle East. Iran may well be on the way to becoming a member of the nuclear states' club—a development that would destabilize the region and potentially lead to a spiraling arms race. Internally, Iran is in a state of great motion, with growing opposition to the current regime—a situation that carries a high imploding potential. The specter of the closure of the Strait of Hormuz is always present. The strait, which assures the flow of 20% of world oil every day, has a very critical place in the world's energy infrastructure. Its blockage would dramatically increase oil prices, and create a real problem for the supply industry, which would have to reroute considerable amounts of oil.

Other potential sources of trouble include Iraq, Egypt and Yemen. Iraq is in the throes of ethnic violence, and its plans to increase oil production in the last two years have not gone smoothly. Egypt does not have important oil resources but is the host of critical supply infrastructure. Both Egypt and Yemen have suffered a number of terrorist attacks in the past few years. In general, the domestic situation in most Middle Eastern countries is very volatile and will probably remain so for some time. Most of these countries introduced a variety of socio-economic reforms in the past year, following the so-called Arab spring of 2011, which brought down long-standing totalitarian leaders. And while the reforms and the changes of leadership were welcome both inside and outside these countries, a rocky path of democratization and development still lies ahead. How this process will evolve is highly uncertain.

An inter-state war in the Middle East is not very likely at this point. However, any of these potential "bombs" might explode at any moment, with dire consequences for the world of oil. The Middle East supplies 24.59% of European crude oil, 28% of US oil and 88.72% of Japan's.[6] An interruption of the flow of Middle Eastern oil would have very adverse effects for industrialized countries.

Disruptions involving oil transit choke points and tanker accidents are another important factor to take into account. A number of critical oil routes transit through potentially volatile regions. In addition to the Strait of Hormuz, the straits of Malacca (located between Malaysia, Indonesia and Singapore), and Bab el-Mandab (located between Yemen, Djibouti and Eritrea) present a cause for concern. These waterways have been plagued by piracy and terrorism for years. European oil supply, in particular, is very vulnerable to disturbances in the Bosporus/Turkish straits and the Suez canal/Sumed pipeline choke points. The Bosporus is a critical link in the transportation of Caspian oil to Europe, as the Suez-Sumed system is in the transportation of Saudi oil. In 2006, about 2.6 mbd of Saudi oil were transported through the pipeline, most of them bound for Europe.[7] Disruptions of oil flows might also occur as a result of natural disasters. Their frequency and intensity has increased in the past decade. An example of this is Hurricane Katrina, which for a short time in 2005, took all the oil rigs and refineries on the Gulf Coast out of operation.

Tanker and oil field accidents are another major cause for concern. While their number has significantly decreased in the last several decades due to stricter regulations, they still occur frequently. One of the biggest ones—the *Exxon-Valdez* accident in 1989—led to the loss of 11 million gallons of oil. Incidents like that have a strong impact on oil prices.

A third factor directly impacting European energy security is the increased reliance on Russian oil—a very attractive source, because of its proximity. Currently, Russia supplies more than 33% of EU consumption—an increase from only 16% in 2001. The change is dramatic and will deeply affect the future of EU-Russia relations. It also has implications for European oil supply security, as oil becomes increasingly concentrated.

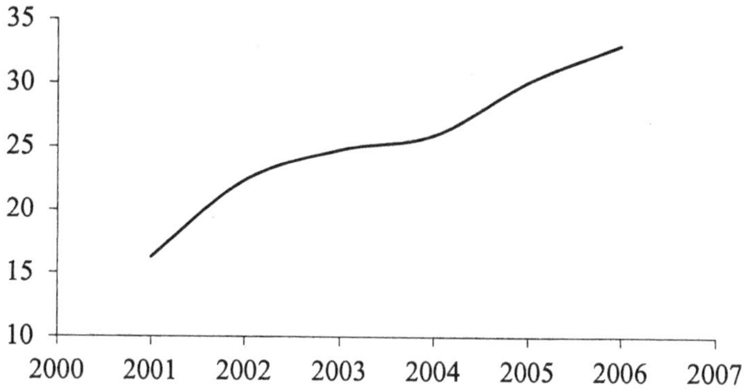

Figure 1. Russian oil imports to the EU (% of all imports)
Source: Data from European Commission, "EU Crude Oil Imports" http://ec.europa.eu/energy/observatory/oil/import_export_en.htm (accessed 10 March 2012)

And lastly, oil will remain the dominant energy source for the foreseeable future. Projections put it at between 32% and 38% of the world's energy consumption in 2030. This is especially relevant for the transportation sector where there are fewer alternatives. These estimates, coupled with the fact that oil is a finite resource, have important implications for the future of oil security. The optimists have often disputed the finiteness of oil, as we keep improving the technology, which allows us to more effectively explore old fields and discover new ones. The rate of new discoveries of oil is very difficult to predict as it depends on technological developments. It is true that 2009 has been one of the most abundant years in recent decades with respect to new discoveries of oil, however, despite the existence of a variety of optimistic projections, fossil fuels are slowly being depleted. In addition, most new oil fields are very expensive to explore. This has led to growing investment in the study and development of renewable energy sources. Their share of energy use is expected to grow at a rate of 3% per year until 2030; however they do not offer a viable alternative to oil, especially in the transportation sector.[8] Oil will remain an essential commodity, at the mercy of the vagaries of the market and increasingly concentrated in a handful of unreliable regimes.

Oil Crises and International Cooperation

All this is to say that energy supply shortfalls are not only incidents of the past. The risk of a supply disruption still exists. Whether it is triggered by the policies of an exporter government, a terrorist attack, disturbances in the Middle East, a natural disaster or by a natural reduction in production, an energy crisis would greatly disrupt the economies of both developed and developing nations. Its consequences would be even more severe if the affected consumer countries were to fail to cooperate in resolving the crisis, as has been the case several times in the past.

International cooperation is a critical factor in any type of crisis. It is especially important, however, when it comes to a finite, highly concentrated and critical commodity like oil. The nature of oil supply resembles a zero-sum game, that is, increasing the supplies of one side inevitably decreases the supplies of another. This has important implications for cooperative behavior especially in times of oil shortages. It might lead to scrambling and oil hoarding if the actors involved do not act cooperatively. And although it is widely known that in cases of zero-sum games cooperation leads to Pareto-optimal solutions, governments seem to forget that when a crisis strikes. In addition, the oil market is highly interdependent, with a variety of actors—governments and private entities—interacting on a daily basis. If a shortage is to be countered effectively, coordination is necessary.

Cooperation among Europeans, between Europeans and other consumers, and between consumers and producers, is essential for coping with oil supply uncertainty. And although this is widely known, even European governments, which have pooled sovereignty on many other issues, find it difficult to coordinate policies on these matters. A variety of initiatives, aimed at increasing the opportunities for dialogue have been already taken. The most notable one is the International Energy Agency (IEA), established in 1974, as a result of the 1973 oil shortage. There are also the International Energy Forum, the Energy Charter Treaty, and a variety of other regional and multilateral organizations and instruments. The activities of most of these bodies are, however, limited to the collection of information.

This study aims to shed light on the issue of consumer countries' cooperation in times of oil crises. The analysis provides a test for two important hypotheses, regarding the conditions under which cooperation is more likely to occur: the institutionalist hypothesis and the hegemonic stability hypothesis. I use the method of structured focused comparison to evaluate European cooperative efforts during seven post–World War II oil supply shortages. The three most important European consumers and importers—Great Britain, France and Germany—will receive particular attention. The cases will be examined in relation to the dominant oil power at the time—the United States—and the existing international institutions.

This study fills several significant gaps in the international relations literature. It evaluates inter-state cooperation in an issue-area that has been largely understudied, and that is becoming increasingly important for international relations. Past events have shown that energy crises can have devastating consequences for the world economy. An oil supply disruption of the future might be even more destructive than it was in the 1970s, due to the higher level of economic interdependence, increased reliance on fossil fuels or the fact that oil reserves are diminishing. The study of consumer country cooperation in this issue-area is therefore, very much overdue, especially given the lack of cooperation in the past and the high gains that could be obtained from it. The issue-area of oil security is important both for scholars of international relations and for practitioners. Oil is an essential ingredient of the modern industrialized economy. It is widely used not only in transportation but also in a variety of industrial processes, such as the production of plastics, fertilizers and medicines. Regular and secure oil supplies are an issue of state survival. Therefore, the analysis of consumers' behavior in times of supply shortages might have important implications for policy makers.

A Dearth of Scholarship

The world has witnessed seven major oil supply crises since 1951, three of which have had dramatic consequences for the world economy. Yet international cooperation during times of energy crises has been the subject of only a few international relations studies. Moreover, none of these studies provides a systematic analysis of all major disruptions. They are limited in their scope and findings. In general, oil has been predominantly the focus of economics. The vagaries of its supply and demand, and the intricate laws of the market, in general, have been given ample attention. The field of international relations has only recently begun to examine oil and its role in inter-state relations, focusing mainly on its use as a foreign policy tool, and Middle Eastern politics. The majority of the literature also concentrates on the United States and its energy policy, while European policy remains on the sideline. Lately, energy security, defined in a variety of ways, has also started featuring in the literature. It is astonishing that very few of these studies have concentrated on relations among consumer countries, even though these form an integral part of consumers' energy security. This is especially true for periods of supply shortages and other market disturbances. It is an established fact that cooperation can mitigate the distribution problems arising from crises of this type. Yet, it has not received much attention by international relations scholars. Furthermore, even though institutionalism has been gaining increasing acceptance in the international relations field, little study has been done on the role of institutions in facilitating cooperation in times of energy shortages. In general, there is a dearth of research on the role of international organizations (IOs) in the energy issue-area.

There exist very few analyses of cooperation in oil. In a series of studies on the 1970s oil crises, Robert Lieber examines the lack of cooperation among consumer countries in 1973 and 1979.[9] The author looks at the cooperative efforts of Western countries (including Japan) in three distinct situations: nonemergency periods, supply shortages below the 7% IEA threshold, and full-fledged crises. His analysis points to the tendency of countries to act unilaterally, when faced with a crisis. Cooperation was at its weakest during the 1973 oil disruption; it was somewhat more prevalent during the next crisis in 1979, which suggests the existence of a learning curve. According to Lieber, the failure of consumer countries to cooperate was due to the following factors: export competition, insecurity and lack of information concerning how long the supply shortage would last, differences between Western energy producers and consumers (e.g., the UK and France), divergent ideas about energy security, and institutional obstacles.[10] He concludes that unilateralist responses are inadequate in a world of high interdependence.

Another important study of consumer countries relations is provided by Ethan Kapstein. He looks at NATO members' cooperation and conflict during six post–World War II energy crises.[11] He argues that alliance relations in the energy issue-area could best be accounted for by two variables—distribution of power, that is, the role of the hegemon, and the allies' hierarchy of interests. In those cases where there were considerable hegemonic resources and allies' interests overlapped, cooperation ensued. In those cases where both of these things were missing, there was a resource scramble. In all other cases, attempts at coordination were made but were not very successful.

These are the only studies that purposefully look at consumer countries' relations in times of energy shortages. Some scholars, while not examining the matter in detail, have mentioned the role of cooperation in regard to energy security. David Deese, for example, argues that international cooperation, "especially in oil allocation systems" can help improve energy security for oil importing countries.[12] Oil, the author continues, is more important to national security than any other commodity. In a study published in 1980, Joseph Nye argues that differing energy vulnerabilities among Western countries will lead to cleavages among them in the future.[13] He also suggests ways to improve cooperation.

The studies mentioned above do not provide a satisfactory understanding of energy crisis cooperation. They are limited in their scope: Lieber only studies the two 1970s disruptions, and Kapstein, while covering a greater number of energy crises, concentrates on relations within NATO. Moreover, the power variable is not sufficiently studied by either of the authors. A comprehensive study of all post–World War II oil supply disruptions does not exist, and there is a pressing need for one, given the increasing importance of energy issues. It is also critical for IR theory development.

The Theoretical Dilemma

Inter-state cooperation has been one of the richest sources of theorizing in international relations. Interestingly, while theories of cooperation have been applied to a variety of issue-areas, oil and energy have generally been omitted. I believe that it is time to remedy this situation. Applying theories to a single very important area of international life provides a valuable opportunity to test these theories' assumptions. In addition, by pitting two of the most important theoretical perspectives in international relations—neorealism and institutionalism—against each other, I perform a kind of a "crucial test."[14] According to Stinchcombe, theories gain credibility by being compared with each other in crucial experiments: "by eliminating the most likely alternative theory, we increase the credibility of our theory much more than if we eliminate alternatives at random."[15] In this study, I compare neorealism and institutionalism by testing their assumptions about international cooperation. By means of logical deduction, empirical statements are derived for both theories, which are then tested in seven cases of oil supply disruptions. Here is a short overview of the theoretical debate.

Both neorealism and neoliberalism have extensively explored the issue of inter-state cooperation. The latter is usually defined as the adjustment of actors' behavior "to the actual or anticipated preferences of others, through a process of policy coordination." Structural realists maintain that anarchy limits the willingness and abilities of actors to cooperate. States cooperate only to counterbalance a concentration of power or advance their own interests. Power is the guiding force in international relations, according to neorealists. Thus, international institutions are effective only as long as they are supported by the dominant actors of the day. Examples abound, neorealists would argue: the World Bank and the International Monetary Fund, established after the Second World War, are examples of effective institutions, supported by the dominant actor, while the periods of stagnation in European integration are examples of an institution's lack of effectiveness, due to the absence of a leading nation. These assumptions have developed on the basis of the hegemonic stability theory, which studies the role of dominant actors in providing leadership and disciplining markets. The hegemon, it is said, has considerable resources and can provide incentives for other countries to cooperate. Though it is based on the hegemonic stability thesis, this study prefers the use of the terms *leadership, power or dominant power.* While very similar in meaning to the term hegemony, these other terms avoid the ambiguities associated with it. In the case of the oil issue-area, a leader would be a country that has a large resource base, that can influence and discipline the market, and that can provide resources as a supplier of last resort.

Neoliberals, the main opponents of neorealism, assert that "cooperation under anarchy" is easier to achieve and outline a number of conditions that facilitate cooperative agreements.[16] Neoliberal institutionalism maintains that formal international institutions and agreements facilitate inter-state cooperation by providing fora for debate, increasing the flow of information, reducing transaction costs, setting standards and identifying focal points.[17] According to institutionalists, regimes are intervening variables between systemic variables, such as power, or its concentration, on one side and outcomes and behavior on the other. Hence, institutions can alleviate the consequences of anarchy in the international system. This does not mean that institutions will always facilitate cooperation. As Keohane and Martin note institutions are not always effective in this respect.[18] The more important questions, the authors argue, concern the conditions under which institutions are critical in the achievement of cooperation and "how to distinguish the effects of underlying conditions from those of the institutions themselves."[19]

This study attempts to shed light on these issues by examining the role of two factors in cooperative behavior: issue-area leadership and international institutions. In particular, the analysis focuses on their effects in the context of one specific class of events—energy supply shortages. Keohane and Martin note that if a coordinating mechanism does not exist, states would not see the benefits of coordinated action, and would not cooperate.[20] This implies that in the presence of such a mechanism, states will opt for cooperative solutions. Why is it, then, that in the presence of IOs, states did cooperate in some instances of oil shortages and did not in others?

In this study, I examine two hypotheses:

1) Nation-states are more likely to cooperate under the leadership of a dominant power. As neorealists argue, the latter has the resources to coerce others into cooperation, and to serve as a supplier of last resort.
2) International institutions can facilitate cooperation. As institutionalists predict, IOs provide a forum for debate and the exchange of information, decrease transaction costs and increase opportunities for issue-linkage. All these factors make cooperation more likely.

The Empirical Conundrum

All postwar oil crises resulted in the loss of a significant portion of the oil supplies from the market. In some of the cases, scrambling and competitive bidding ensued, exacerbating the crisis. In these cases, a deep and prolonged recession followed, as oil prices were driven up by the competition. In other cases, the crisis had a much smaller effect on the market and the economy in general. This variance seems to be best explained by levels of cooperation, among other factors. In the cases where states coordinated their emergency policies, the crises were milder. In contrast, when cooperation failed, the crises were severe. The question I try to answer is what were the conditions under which states cooperated?

The Cases

Three of the seven cases are examined in detail. The first serious oil supply disruption was caused by the closure of the Suez Canal in 1956. Seventy percent of Middle Eastern oil destined for Europe was transported through the Canal. In addition, Britain and France were embargoed by Saudi Arabia. This resulted in the loss of 2 mbd or 90% of European consumption for a period of four months. This was a major oil shortage, seriously compromising European economic recovery and growth. In relative terms, Suez was the biggest oil shortage in the postwar period. Yet, it had a much smaller impact on consumers, than did later crises. The evidence suggests that the main reason for that was the high degree of coordination in the response to the crisis. European governments, together with the United States and the oil majors, gathered in a number of committees and managed to avert a major economic disaster.

Another severe supply disruption occurred as a result of the Yom Kippur war and the subsequent Arab oil embargo in October 1973. On average 10% of

European oil supplies were cut, and prices increased 400%. This crisis provided the opportunity to test the newly adopted oil emergency procedures. Unfortunately, no emergency coordination took place, and European countries scrambled for supplies, signing bilateral contracts with producers and stockpiling oil. Unilateral action would characterize most of the post-1973 energy shortages. The absence of a united consumers' front allowed Arab producers to raise prices significantly, even though Arab oil cuts were largely made up by production elsewhere.

The decade presented the world with another severe energy crisis, caused by the Iranian revolution. In December 1978, all oil production in Iran halted. Iran, at the time was the second largest exporter of oil in the world after Saudi Arabia, accounting for 10% of noncommunist world consumption. The crisis caused the second major oil price jump in five years. It also presented an opportunity to test the capabilities of the newly created IEA, which featured oil crisis management among its responsibilities. Unfortunately, this higher level of institutionalization did not produce a corresponding level of policy coordination.

The selection of these three cases was based on several important considerations. The 1970s crises represent the two biggest peace time oil disruptions in history. The Suez crisis was chosen, because it was the first major oil disruption, seriously affecting all three countries. The detailed study of these cases also provides the opportunity to test the three institutions under consideration, which had differing institutional structures and powers—the Organization for European Economic Cooperation (OEEC)—Suez Crisis, the Organization for Economic Cooperation and Development (OECD)—1973 crisis, and the International Energy Agency—1979 crisis.

The four relatively smaller crises occurred in 1951, 1967, 1980, and 1990. They are smaller as measured by the impact they had on the world economy and not the actual shortage. In fact, the 1951 oil crisis ranks as the second biggest in terms of percentage shortfall in Europe.

The latest supply shortfall—caused by Hurricane Katrina in the United States—will be very briefly discussed, as well. It is not one of the case studies, however, it deserves some attention, being the only post-1970 crisis, where some level of policy coordination occurred. Its consideration also increases the contemporary relevance of the study.

Table 1 provides a summary of all cases. These numbers have to be treated with caution. They are approximate, as some of these variables are difficult to calculate. They also conceal a more complex truth, which cannot be captured by a single number. For example, in 1951, only aviation fuel was in short supply, and there was plenty of crude oil on the market. These case-specific details will be elaborated on in my discussion of each case. This table is provided only for reference.

Table 1. Major global oil supply disruptions since 1951

Date of Oil Supply Disruption	**Average *Gross* Supply Shortfall (mbd)**	**Supply shortfall in Western Europe (% of consumption)**	**Oil price increase (%)**	**Reason for Oil Supply Disruption**
6/51–9/51	0.7	54		Iranian oil fields nationalized
11/56–3/57	2.0	60	40	Suez Crisis
6/67–8/67	2.0	15		Six Day War
10/73–3/74	2.6	12	400	Arab oil embargo
11/78–4/79	3.5	15	150	Iranian revolution
10/80–12/80	3.3	14.5	150	Outbreak of Iran-Iraq War
08/90–03/91	4.8	32.4	118	First Gulf war
08/05–10/05	2	0	5.6	Hurricane Katrina

The Power Variable

One of the hypotheses advanced in this study is that nation-states are more likely to cooperate under the leadership of a dominant power. Most scholars look at power as a global phenomenon extending across issues. For the purpose of this study, the term issue-area power is used instead. In the case of oil, the dominant nation would have high spare production capacity, a controlling share of world oil production and market influence through oil companies. The United States has been the only actor which has had all three elements at the same time. While there have been other countries with high spare capacity or strong production at different times, they have never been in control of all elements together. This is why US power in oil is considered as one of the independent variables.

The International Organizations in Focus

The second independent variable is international organizations. During the period under discussion, several international institutions attempted to facilitate relations in the oil issue-area. The oil committee of the Organization for European Economic Cooperation was the first international body, assigned with some responsibilities in the field of oil. It did not have emergency authority, as an oil shortage was not expected to occur in peacetime. Its main function was to coordinate oil refinery expansion in Europe, financed in part by the Marshal Plan. Unexpectedly, the oil committee got involved in two oil shortages in the 1950s—the Abadan and the Suez crises. It became clear that the supply of oil could not be taken for granted and that a crisis coordination mechanism was necessary. The role of crisis coordinator was given to the new oil committee, established within the Organization for Economic Cooperation and Development in 1961. This committee fared well in the 1960s, but after the 1973 turmoil, a decision was made to establish a full-fledged energy organization, and the International Energy Agency came into being. It embodied the highest level of institutionalization in the energy issue-area. At the time of the 1979 oil crisis, the Agency had an elaborated energy emergency program, and was responsible for establishing an energy market information system, while also promoting multilateral energy research and development projects. In some of the cases, the role of the European Economic Community (EEC) will be considered, as well. The EEC adopted its first oil emergency measures in 1968, as a result of the June 1967 embargo. Therefore, its role in coordinating crisis responses is evaluated only for the cases after 1970.

The theoretical importance of this study is, without a doubt, unquestionable, as it contrasts the cooperation propositions of two mainstream international relations approaches. While the field has witnessed a rich theoretical diversification in the 1980s and 1990s with the advent of alternative theoretical perspectives, such as constructivism, the two *neos* remain dominant. Let me turn next to the policy implications of this analysis.

Oil Prices: Why Do We Care?

A significant price increase is one of the most serious consequences of oil shortages. The price of oil affects prices in general, and in turn unemployment and GDP growth. A dramatic oil price rise can lead to a recession, as was the case with several of the post-1970 oil crises (see figure 2).[21] For example, after the first oil shock in 1973, unemployment in the United States reached 8.5%—its highest level since the Second World War. In Europe, inflation was in the double digits throughout the 1970s. The 1980 and 1990 crises had similar consequences, though smaller in magnitude. In the 2000s, for the first time prices increased not as a result of an oil shortage, but a combination of factors, including strong demand growth and low spare capacity. Whatever the cause, the consequences of an oil price increase for the world economy are significant. I focus here, however, only on the effects of oil supply shortages.

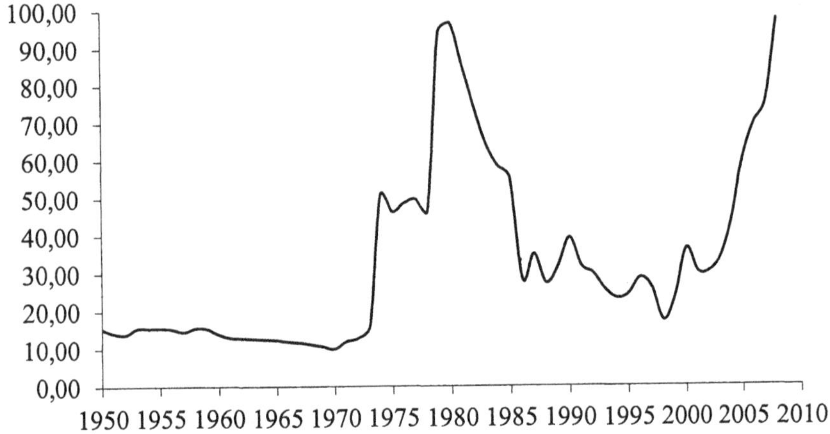

Figure 2. Oil prices, 1950–2009
Source: Data from British Petroleum, "Statistical Review of World Energy" http://www.bp.com/productlanding.do?categoryId=6929&contentId=7044622 (accessed 15 January 2010)

Most of the crises that have occurred in the post–World War II era were psychological, and affected expectations rather than the physical supply of oil. Some authors even speak of two types of energy vulnerability: physical supply vulnerability and economic vulnerability.[22] The first refers to the risk of a physical interruption of oil supplies, and the second—to the risk of high oil prices and their impact on the economy. I argue that the second risk is much more important than the first, in which case policy coordination becomes an important variable. In most cases (with the possible exception of the 1973 oil embargo), the oil cut was offset by increased production elsewhere. For example, in 1980, when Iranian and Iraqi oil production were removed from the world market, increased production from Saudi Arabia, Kuwait, Norway and Britain more than offset the shortfall, which actually led to an oversupply in early 1981. Oil prices, however, increased with almost 150%. The increase was fueled by uncertainty as to the duration of the shortage, and the impact of the Iran-Iraq war on the region. Market participants were fearful of the possible spread of the conflict to Kuwait and Saudi Arabia, which could remove most Middle Eastern oil production from world markets. Naturally, this uncertainty led to panic buying and hoarding of oil—an extremely counterproductive behavior, which exacerbated the crisis, by further driving the price up.

In sum, in most cases of oil supply disruptions, prices increased not because of the actual shortage of oil, but because of the self-reinforcing effects of two factors: uncertainty about the future, and the consequent panic buying. These effects were much more prominent in the crises occurring after 1970. This variance is best explained by consumer cooperation, which considerably reduced the level of uncertainty. Therefore, the question I seek to answer is why is it that European consumers found it much easier to cooperate in some cases than in others?

To answer this, I look at two important variables: issue-area leadership and international organizations. The single most important reason why these two factors are selected is that they are both offered as the best explanations by the two contending theories. Scientific rigor, however, requires the control of the eventual influence of third variables. In this case, candidates for alternative explanations are oil market conditions at the time of the crisis, and the structure of the market, in general.

Oil Market Conditions

Conditions on the oil market include three important variables, which could either aggravate or alleviate a crisis. These are world spare capacity, inventory levels, and the share of oil in the energy mix. Depending on their degree, these factors can considerably worsen or mitigate the consequences of a supply shortage. Some of them also provide a security cushion against shortfalls, which might have an effect on the prospects for cooperation.

Spare capacity indicates the excess production capacity that can be put in operation in a very short period of time. It provides a sort of an oil shortage hedge, as excess production in one country could be put in operation, in the case of a disruption from another source. It has also proven to be an important determinant of oil prices. In the mid-2000s—due to exceptionally high oil demand in Asia—spare capacity reached its lowest level since the early 1970s. This drove prices up 122% between 2004 and 2006. Ideally, spare capacity would provide a psychological buffer against supply disruptions and would increase the prospects for cooperation. In this study, I look at American spare capacity, as I expect it to have a particularly strong impact on trans-Atlantic and European cooperation. The dynamic again involves factors of perception. American excess capacity would be of special importance for American allies, more so than spare capacity held by any other country. It would have the added value of being in "friendly" hands. Western Europe's energy security would be more robust if its determinants were in allied control. In addition, it is considered here as part of a combination of factors that make up one of the independent variables—oil power. No other country has possessed both elements of power in the oil issue-area, that is, control over resources and the market. This is why US power in oil is considered separately. Since world spare capacity might have a non-negligible influence on oil market developments, it is also discussed for some of the cases.

Levels of oil stocks held by both private and public actors in consumer countries could also provide an important security cushion. An international stock-holding requirement was introduced for the first time by the OECD oil committee in 1962, as part of its emergency program. Member-states were required to hold stocks equivalent to 60 days of consumption. Several years later, stock-holding measures were adopted within the EEC, as well. After the experience of several oil shortages in the 1970s, the requirement was increased to 90 days and inventories came to be seen as the primary shield against oil supply disruptions. As of September 2010, OECD countries held 4.3 billion barrels of oil stocks.[23] Ideally, they should be re-filled during times of normalcy, and used in times of shortage. Their use has two purposes—offsetting the shortfall and calming the market. As they are considered an important part of the international oil infrastructure, the levels of oil stocks will also be discussed.

The third factor—level of oil dependency—might seem relevant at a first glance as European dependence on oil has gradually increased in the post–World War II period. The difference is indeed striking. In 1951, OECD-Europe relied on oil for 18% of its energy needs, and in 1970—60%. This tremendous increase was due to a variety of factors, including strong economic growth, and fuel-switching from coal. Critics would argue that post-1970 crises were more severe simply because oil became a more critical commodity during the 1960s. An interruption of oil supplies had much more serious consequences after 1970 than before. An examination of two cases with a great variance on the dependent variable—cooperation—reveals a different story. In 1967 and 1973 European reliance on oil was at similar levels, yet the levels of cooperation greatly differed. Thus, oil dependency does not have a strong explanatory value. Currently, oil makes up 35% of global primary energy consumption and its share keeps declining. This has implications for the possible effects of oil shortages on the overall economy, but not so much for cooperation.

Spare capacity, inventory levels, and oil dependency can have a mitigating crisis effect. The first two are more important as they are more vulnerable to changes in the short-term. Conditions of high excess capacity and high stock levels slightly lower crisis uncertainty and might increase the chances for cooperation. I call these favorable market conditions, and examine their role in most crises.

Market Structure

The second factor—market structure—has a more important influence on the studied variable. The oil market underwent significant changes throughout the studied period. After the Second World War, production and trade were in the hands of the oil majors, the so-called seven sisters—Standard Oil of New Jersey (later Exxon), Standard Oil of New York (later Mobil), Standard Oil of California (later Shevron), Gulf Oil and Texaco (both merged with Shevron), Anglo-Persian Oil Company (later BP), and Shell. Vertically integrated, these companies owned the entire oil infrastructure—from the oil fields to the gas pump.

Later, smaller companies, the independents, entered the market, partially as a result of American antitrust legislation. But the biggest change came in the late 1960s and early 1970s: producers' emancipation. One after the other, producer governments took control over their oil fields, depriving the companies of an important lever. It is important to note that these companies performed the critical function of redistributing available supplies in times of shortage.

Other major transformations of the oil market include price deregulations in the 1980s and the advent of spot trading, as opposed to long-term contracts. In 1983 the New York Mercantile Exchange started trading oil futures, providing an important hedge against price risk and volatility. The market became a sort of an independent entity, capable of absorbing smaller shocks. This was a big departure from a regulated market, dominated by long-term agreements.

Do these changes affect oil crises cooperation? Market forces play an important adjustment role, naturally balancing supply and demand in response to price fluctuations. Oil supply and demand, however, are relatively inelastic to price changes. It takes years before demand patterns change in response to high oil prices, and in turn, bring adjustments to the supply side. In the meantime, high oil prices bring high economic and social costs, in the form of high inflation and increased unemployment. By coordinating action during supply emergencies, consumers could reduce price volatility risk and mitigate its negative consequences. In addition, while market forces might be able to absorb smaller shocks, they might also make cooperation harder to achieve, simply because the number of actors and the complexity of the market have increased.

The factors, driving cooperative behavior, are mostly of political nature and remain the same. Oil is a commodity traded like any other. Its significance for national security, however, distinguishes it from other commodities and introduces political factors into its analysis. In fact, the politics comes to dominate the economics, in the case of oil. I believe inter-state cooperation is a function of factors lying outside of the economic sphere. However, the market environment did have some minimal impact on the outcome of the crisis and will be controlled for as much as possible. For this purpose, variables like demand-to-production ratio and commercial stocks are reported, as well.

These additional factors, while briefly mentioned, where relevant, are kept to a minimum, in order to keep the model parsimonious and simple. The process of modeling social relationships always requires simplifications, and as such, reflects reality only to a degree. Simplification, however, cannot be avoided, especially with regard to the complex phenomena studied in international relations. This study still helps increase our understanding of the dynamics of cooperation in times of oil supply shortages, by outlining the most important conditions under which it is more likely to occur.

The Method of Structured, Focused Comparison

Comparing historical cases is always a controversial matter. What factors should we consider among the myriad of details? How do we distinguish what is important among all the historical "noise"? How do we treat the idiosyncratic elements of the particular point of time? To minimize these difficulties, I use the method of structured focused comparison, developed by Alexander George, which enables the systematic study of historical data.[24] This method involves a two-step process: (1) the selection of cases, belonging to a particular class of phenomena, and (2) the formulation of a set of general questions to be used in each case. This process makes it easier for cases to be considered outside of their particular historical context, and generalizable lessons to be drawn. It is a very useful methodology, allowing the rigorous testing of political theories through historical analysis.

The application of the structured-focused method requires the comparison of identical elements across a number of different cases. In this volume, seven cases of past energy crises will be surveyed (see table 1), and the emergency responses of Great Britain, France and Germany will receive particular attention, since they are the biggest oil consumers in Europe, and the dominant actors in European politics, in general. For each case, I will seek to answer the following questions:

- How did the United States administration respond to the crisis? What was its immediate reaction and how long did it take to implement any crisis-related decisions?
- How did the French, German and British governments respond to the crisis?
- Did any policy coordination among the above governments (and European states, in general) take place within the respective international institution? A cooperative policy response includes coordinated draw-down of petroleum stocks, the redistribution of oil to countries experiencing the biggest shortfall, measures for alleviating competitive spot market bidding and a coordinated policy toward exporting countries.
- Did the competencies and responsibilities of the respective institution have any impact on the decisions taken?

In addition, as a matter of statistical evidence—a number of indicators will be reported for each case. US oil resources, as an indicator of a power status, are the most important variable. European dependence on the interrupted oil source, and the magnitude of the shortage will also be taken into account.

The study will focus on the period during and immediately after the crisis. I will look for evidence of policy coordination in both primary and secondary sources. Two types of primary sources are used: documents of the studied organizations, such as memoranda, briefings and correspondence, and documents originating from the respective national administrations, in particular the United States. The US government resources are selected primarily because of the ease of access to them, but also because of the important role the United States and the American oil majors played in most oil disruptions. The secondary sources are mostly historical accounts of the events. Press coverage will also play an important role in my evaluation of the available evidence. The following newspapers and magazines have been consulted for most cases: The *New York Times*,

Le Monde, *Frankfurter Allgemeine Zeitung*, and *The Economist*. For later cases the Oil&Gas Journal and the Petroleum Economist have also been examined.

The structural nature of this case study comparison does not preclude the evaluation, in some of the cases, of idiosyncratic factors, such as the dynamic of the US-Soviet rivalry at the time of the crisis. In a case study, covering such a long period of time, it is impossible to completely account for all time-related factors. For example, the 1950s saw the beginning of the demise of European influence in the Middle East and the advent of US predominance. The 1970s were a particularly turbulent period of time, which witnessed the beginning of collective action in the Third World, the end of the Vietnam War, the opening of China, and the Egypt-Israeli accords, among other events. Nineteen ninety—the year the Gulf crisis started—was the first year of the post–Cold War era—the world was still baffled by the magnitude of the events, and the insecurity that ensued. To the extent possible, these factors will be taken into account.

The next chapter provides an overview of the theoretical debate between neorealism and neoliberalism.

Notes

1. Organization for European Economic Cooperation, Report by the Oil Committee to Be Published in 1956 (Paris: OEEC, 1956), 3.

2. The Organization of Petroleum Exporting Countries currently has twelve members: Algeria, Andorra, Ecuador, Iran, Iraq, Kuwait, Libya, Nigeria, Qatar, Saudi Arabia, United Arab Emirates and Venezuela.

3. U.S. Energy Information Administration, "Short-term Energy Outlook: February 10, 2010 Release," http://www.eia.doe.gov/pub/forecasting/steo/oldsteos/feb10.pdf (accessed 15 February 2010).

4. See, for example, Christopher M. Blanchard and Alfred Prados, "Saudi Arabia Terrorist Financing Issues," Congressional Research Service, http://www.fas.org/sgp/crs/mideast/RL32499.pdf (accessed 16 February 2011).

5. U.S. Energy Information Administration, "World Proved Reserves of Oil, Most Recent Estimates," http://www.eia.doe.gov/emeu/international/reserves.html (accessed 15 February 2011).

6. International Energy Agency, *Oil Information 2002: Donnees Sur Le Petrole* (Paris: IEA, 2002).

7. U.S. Energy Information Administration, "World Oil Transit Chokepoints: Suez/Sumed," http://www.eia.doe.gov/cabs/World_Oil_Transit_Chokepoints/Suez.html (accessed 20 February 2011).

8. U.S. Energy Information Administration, "International Energy Outlook," http://www.eia.doe.gov/oiaf/ieo/index.html (accessed 15 February 2011).

9. Robert J. Lieber, *The Oil Decade: Conflict and Cooperation in the West* (New York: Praeger Publishers, 1983).

10. Ibid., 11.

11. Ethan Kapstein, *The Insecure Alliance: Energy Crises and Western Politics Since 1944* (New York: Oxford University Press, 1990).

12. David A. Deese, "Energy: Economics, Politics, and Security," *International Security* 4, no. 3 (1979): 7.

13. Joseph Nye, "Energy Nightmares," *Foreign Policy*, 1980.

14. Arthur L. Stinchcombe, *Constructing Social Theories* (New York: Harcourt, Brace & World, 1968), 25.

15. Ibid.

16. Kenneth A. Oye, *Cooperation Under Anarchy* (Princeton, NJ: Princeton University Press, 1986).

17. Robert Keohane and Lisa Martin, "Institutional Theory as a Research Program," in *Progress in International Relations Theory: Appraising the Field*, ed. Colin Elman and Miriam Fendius Elman (Cambridge, MA: MIT Press, 2003), 80.

18. Robert O. Keohane and Lisa L. Martin, "The Promise of Institutionalist Theory," *International Security* 20, no. 1 (1995): 47.

19. Ibid.

20. Ibid.: 45.

21. For a detailed discussion on the effects of oil price increases on the economy, see Robert B. Barsky and Lutz Kilian, *Oil and the Macroeconomy Since the 1970s* (Cambridge, MA: National Bureau of Economic Research, 2004); Pascal Ditte and Peter Roell, "Past Oil Price Shocks: Political Background and Economic Impact," *ISPSW Publications* (2006), http://www.isn.ethz.ch/isn/DigitalLibrary /Publications/Detail/?id=20499 (accessed 23 February, 2011); James D. Hamilton, "Oil and the Macroeconomy Since World War II," *The Journal of Political Economy* 91, no. 2 (1983); International Energy Agency, "Analysis of the Impact of High Oil Prices on the Global Economy," (2004), http://www.iea.org/papers/2004/high_oil_prices.pdf (accessed 23 February 2011); Lutz Kilian, "Exogenous Oil Supply Shocks: How Big Are They and How Much Do They Matter for the US Economy?," *CEPR Discussion Papers* (2006), http://bankofcanada.ca/en/conference_papers/commodity_price2006/kilian.pdf (accessed 23 February 2011); Martin Schneider, "The Impact of Oil Price Changes on Growth and Inflation," *Monetary Policy and the Economy*, no. 2 (2004), http://www.oenb.at/en/img/mop_20042_ the_impact_tcm16-19678.pdf (accessed 23 February, 2011).

22. Robbert Willenborg, Christophe Tonjes, and Wilbur Perlot, "Europe's Oil Defences: An Analysis of Europe's Oil Supply Vulnerability and its Emergency Oil Stockholding Systems," (2004), http://www.clingendael.nl/publications/2004/20040100_ciep_paper_willenborg.pdf (accessed 24 February, 2011).

23. U.S. Energy Information Administration, "International Petroleum Stocks," http: //www.eia.doe.gov/emeu/international/oilstocks.html (accessed 30 September, 2011).

24. Alexander L. George, "Case Studies and Theory Development: The Method of Structured, Focused Comparison," in *Diplomacy: New Approaches in Theory, History, and Policy*, ed. Paul Gordon Lauren (New York: Free Press, 1979).

Chapter 2
The Theoretical Debate

The study of international cooperation has long attracted scholars' attention and has generated a large body of literature. With the advent of economic and game theory apparatus, it became clear that, if achievable, cooperative behavior leads to win-win situations. This finding generated a lot of interest in the international relations community, which led to the emergence of one of the most productive debates in the literature—the debate between neorealists and neoliberals.

In the 1970s, the debate expanded into looking at international organizations and their role in fostering cooperative behavior. This turn was nourished by the spread of international institutions after the Second World War. The United Nations and its affiliated organizations gave new hopes for enduring peace, and the Bretton Woods institutions provided for a stable economic order. Later on, the European Community and a number of other regional agreements came into being. All these had one fundamental goal—to find cooperative solutions to common problems. While they have had varying degrees of success, the most accomplished one has been and remains the European Union. However, while it has been a beacon of cooperation in some matters, the EU has seen discord in others. For example, membership in the Union has had a strong effect on cooperative behavior in the so-called low politics areas, and a weaker one in the high politics realms of foreign policy and security. The United Nations organization has seen a similar development—relatively successful in certain technical fields such as public health, but not very productive in areas such as peace and security.

Many international organizations have shared the lot of the EU and the UN. They have been successful at fostering cooperation in some cases and not in others, or in some areas and not in others. This phenomenon has given rise to a new debate in the cooperation literature, dealing with the question of whether and under what conditions institutions foster cooperative behavior. The debate was especially productive during the 1980s and 1990s, while slightly flagging in the 2000s, mainly due to a shift in scholarly attention to issues such as terrorism and Muslim politics. The question of whether and how institutions foster cooperation, however, remains critical in both the empirical and the theoretical sense. Institutions continue to play an important role in international politics. Their utility has exhibited variations over time, due to changing balance of power configurations and other realities. Nevertheless, they endure as an important forum for cooperative behavior, the necessity of which is undiminished.

This study aims at contributing to this grand debate in international relations theory by looking at cooperative behavior in the oil issue-area. The economics of oil has been studied in detail by both economists and political scientists. The effects of oil shortages on the economy, especially as they relate to the onset of recession have been thoroughly examined and understood. It is a well-known fact that these effects could be mitigated if consumers worked together during an oil shortage. Consumers' cooperation, however, has not been granted sufficient attention. This is surprising, given that cooperative action not only alleviates a crisis's consequences, but also impacts consumer-producer relations and supply security, in general. Coordinated action, for example, could prevent or alleviate the price hikes, produced by competitive overbidding. Such a price increase was, to a great degree, responsible for the recession in the Western world in the 1970s.

Studying cooperation in the oil issue-area is important from a theoretical point of view, as well. The supply of oil is a crucial area of inter-state relations, involving both economic and security issues. Thus, it provides a critical test for the discussed theories. In addition, comparing the two most important IR perspectives allows me to perform a sort of a "crucial test."[1] Stinchcombe argued that the best test of a theory, the so called crucial experiment, occurs when the theory is pitted against its main opponent. By eliminating the most likely alternative, the theory tested gains much more credibility.

The findings of this study confirm that when consumers achieved a higher level of coordination, the crisis was less severe. In cases where serious coordination problems persisted—for example, in 1973 and 1979—the outcome was different. It is, therefore, critical, from both theoretical and empirical points of view, to examine, what factors are conducive to cooperative behavior.

Defining Cooperation

Before exploring the neorealist-institutionalist debate, a few definitions are necessary. Although quite complex and versatile, the terms *international cooperation* and *international institution* have been clearly defined in the literature. It is generally accepted that "cooperation occurs when actors adjust their behavior to the actual or anticipated preferences of others through a process of policy coordination."[2] Kenneth Oye adds that cooperation has to be a "conscious" political process.[3] This definition does not imply policy coordination within an established institution with headquarters and staff. Bilateral or other forms of multilateral cooperation are still very prevalent, and, in many cases, the mode of operation preferred by state leaders. The Concert of Europe, for example, which was the backbone of European relations in the 19th century, was never institutionalized. Similarly, in the area of oil, the "Seven Sisters" effectively managed an informal regime until oil assets were nationalized by producers in the 1960s and 1970s.[4]

My focus, however, will be on established international institutions and their effect on state behavior. The terms *international institution* and *international organization* will be used interchangeably in this study. I will also, when the necessary information is available, discuss bilateral interactions, as they relate, indirectly, to the main assumption of this study. In other words, states' preference for bilateral deals in lieu of institutionalized cooperation in time of little or no leadership in oil would be a corroboration of my hypothesis.

The three institutions discussed in this volume—the OEEC and the OECD oil committees, and later the IEA—were until recently the only institutions set to organize states' relations in the oil issue-area. The OEEC represented a looser regime, providing an institutional setting for the exchange of information and for coordination of certain policies, such as refinery expansion. The OECD had more clear-cut rules in place, including a stock requirement. The IEA provides the strongest and most binding set of procedures. In this regard, there has been an evolution toward stronger institutionalization, provoked mainly by the recurrent oil shortages of the post–World War II era, and the realization that coordination of policies offers the best approach.

The Debate

This study attempts to test the claim, advanced by institutionalist theory in response to neorealism, that international institutions foster cooperative behavior. These two theoretical schools have produced a voluminous literature on various aspects of international politics. My focus is on their respective views on inter-state cooperation and international institutions. While they have similar understandings of the nature of international politics in general, they disagree on the prevalence and effects of inter-state cooperation. The similarity of their assumptions has prompted some scholars to study them together, in opposition to the rest of IR theory.[5]

Neorealism and neoliberalism share the following fundamental assumptions:

- The international system is anarchic
- Nation-states are the primary actors in world politics
- States are rational actors, seeking to maximize power
- States are mainly preoccupied with their survival.

Both schools agree that it is difficult to achieve cooperation in this state of affairs. In a system defined by anarchy, self-help is the dominant choice of action. International institutions can help only marginally. Unlike neorealists, however, neoliberals believe that there exists an "unrealized or potential cooperation" in the international system.[6] One of the tools to foster cooperation, they believe, is institutionalization. Through a variety of mechanisms, international organizations help solve coordination problems, such as cheating, and in this way make cooperation a preferred option for states.

Thus, the central points of contention between the two schools concern the opportunities for cooperation and the role of international institutions. According to neorealists, states will cooperate only to further their own interests, and respectively, will not cooperate, if these are threatened.[7] This is a direct consequence of the anarchic nature of the international system. In the absence of an overarching government, if a state's survival is at risk, cooperation is doomed. This is especially relevant in the security field, but could also be applied to other areas. For example, the United States Senate refused to ratify the Comprehensive Test Ban Treaty in 1999 on the grounds of national security. In the political economy field, the protracted and unsuccessful climate change negotiations present a case in point. Restrictions on the fishing of endangered species, which threaten the interests of the fishing industry, have been very difficult to coordinate even within the EU. CO_2 emissions reduction, which could harm developing countries' industrialization efforts, is another example.

Neoliberals agree that inter-state cooperation is hard to achieve. However, they see more opportunities for cooperative behavior. Keohane argues that intensive iterated interaction among a small number of actors can lead to long-term cooperation.[8] Axelrod, using the prisoner's dilemma, finds high reciprocity and iterated interaction as the most important factors promoting cooperative behavior. Hence, he advises leaders to "enlarge the shadow of the future."[9] Oye adds to this list the payoff structure, that is, the benefits of cooperation relative to the benefits of defection.[10]

According to Axelrod, in a prisoner's dilemma setting, the strategy of TIT-FOR-TAT is the most successful in bringing cooperation. In TIT-FOR-TAT, the player cooperates on the first move, and then does whatever the other player did on the previous move. It follows that states, which interact regularly and reciprocate each others' actions, can easily achieve cooperation. This was the case, for example, with the establishment of the postwar economic order, under US leadership.[11] The United States used a TIT-FOR-TAT strategy, giving large concessions initially to build the foundations of the system but relying on reciprocity later. In this way, the United States offered solutions to coordination problems and "promoted relevant conventions of cooperation."[12]

This also means that states might eventually choose to forgo short-term gains in the name of long-term benefits. In fact, not many examples of this type of behavior exist in international relations. The United States' support of the Bretton Woods exchange rate system until 1971 is one of them. Even though it led to an overvalued dollar and balance of payments deficit, the United States supported the system for the sake of a stable international financial order. Similarly, Saudi Arabia played the role of a swing producer in the early 1980s, to support the falling oil prices. This led to a dramatic decline in Saudi revenues, an abandonment of their development program, and a huge budget deficit. The Saudi government discontinued its price support policy in 1985.

These could, however, also serve as examples of states withdrawing from earlier commitments, when those commitments no longer serve their interests. Neorealists would argue that the main reason for this behavior is states' concern for their physical survival, which is dependent on their relative capabilities. In every relationship, states try to augment their power relative to their peers and they do not participate in cooperative agreements that jeopardize their relative position. Hence, Grieco argues, states are "defensive positionalists."[13] Grieco, who also thinks that neorealism is "logically superior" to institutionalism, has found proof of his assumption in the case of the Tokyo round of negotiations on non-tariff barriers to trade.[14] He studied the patterns of EU-US cooperation across different trade issues, such as anti-dumping, government procurement and subsidies, among others. He found that each one of the actors cooperated on the issues, in which they made more gains relative to the other. The pattern of cooperation did not vary according to the number of players involved, the number of iterations or the absolute size of the gains, as institutionalists would have predicted.

Neoliberals see cheating as a bigger obstacle to cooperation than relative capabilities. It is difficult to enforce commitments in an anarchical environment, which leads to mistrust and hinders cooperation. Fortunately, institutionalists argue, there are ways to reduce cheating. International institutions, which could also alleviate the relative gains problem, are one of the mechanisms to do so. This is the second, and most profound, difference between the two theoretical schools—the role of international institutions.

International Institutions

Neorealists see little utility for IOs in the international system. States will participate in institutional arrangements only if these advance their interests, and will opt out of them every time their interests are threatened.[15] France's withdrawal from the military structure of NATO in 1966 and Britain's from the European Exchange Rate Mechanism in 1992 are just two of the many examples of this principle in international politics. For Mearsheimer, institutions merely reflect the balance of power in the international system. A great power, for example, will create and maintain institutional arrangements in a way that preserves and increases its share of world power.

The effectiveness of institutions is also a function of states' willingness to use them as a tool. The establishment of the IMF and the World Bank could be seen in this light. Both were created at the end of the Second World War, when the United States accounted for 50% of the world economy and had a dominant voice in international affairs. It is not surprising then that the United States is the only country in the IMF executive board able to block a supermajority with its own vote, which represents 17% of all votes. European countries, according to many, are also overrepresented, holding one third of the vote. Similarly, the president of the World Bank has always been a citizen of the United States, which is also its biggest shareholder. The five permanent members' veto power in the UN Security Council is an even more prominent example. The inability of states to reform this body in recent decades provides, for neorealists, yet more proof of the power-interest paradigm. What the latter cannot explain, however, is the continued existence of the European Union, where states have transferred sovereignty on certain issues to a higher authority. It is clear that there are winners and losers in the European integration project. A good case in point is the never-ending debate over contributions to the EU budget. Germany, the Netherlands, Sweden and several other MSs are net contributors, meaning that they pay more than they receive. Greece, Portugal and Spain, together with the new members from Eastern Europe, are net receivers. Integration opponents in net contributor countries regularly, and increasingly often, use this as an argument against integration.

Italy's adoption of the Euro provides another example. By adopting the new currency, the government lost an important tool of monetary leverage, one often used to prop up the economy—devaluation of the lira. Indeed, the Italian unemployment rate rose after the introduction of the Euro, and Italy became the first country to seriously consider leaving the euro-zone. Despite the drawbacks, EU member-states manage to stay together and to continue strengthening the bonds that unite them.

Why is it that international institutions prove to be a useful cooperation tool in some cases but not in others? The puzzle is even more intriguing when the same institution is more successful in some cases than in others. This is the case, for example with the OECD oil committee. And this is the question that this study attempts to answer.

While Waltz and his followers are convinced that institutions are only projections of states' power and interests, Keohane and his students think that international institutions play a non-marginal role in fostering cooperation.[16] Institutions play a key role in situations of both shared and conflicting interests on multiple, overlapping issues, where agreement can be reached through bargaining. They can induce states to cooperate by lowering transaction costs; reducing irresponsibility and moral hazard, thereby tackling the cheating problem; providing information, and increasing issue-linkage. In short, institutions help states overcome "political market failure."[17] For example, the increased flow of information generated by institutions can help states increase trust, reduce their fear from each other, and even re-evaluate their interests, leading to a new understanding of the benefits from cooperation. Institutions also create a favorable setting for issue-linkage, which provides new incentives for cooperation and strengthens existing ties. Cases of issue-linkage abound in international politics. Lisa Martin, for example, has found that in the 1982 Falklands crisis, Britain's concessions on other issues gained it the support of all EC members. She found issue-linkage to be the most powerful explanatory factor of cooperation in the case of sanctions. Long-standing institutions also produce a "lock-in" effect, whereby it is costlier for states to leave than to continue. Continued interaction within the same setting also generates more opportunities for cooperation, what functionalists have called a spillover effect.[18] European integration is, again, a good illustration.

Additionally, the ability of actors to distinguish between cooperation and defection is critical for the achievement of long-term cooperative behavior. This can sometimes be difficult, as in the case of arms control agreements. In cases like this, institutions can help by defining cooperative and uncooperative behavior and providing verification mechanisms.[19] In the case of oil, for example, the IEA sets clear rules of cooperation in the event of an oil supply reduction of 7% or above, including provisions for the maintenance of oil reserves, fuel-switching, and a system of oil reallocation if necessary.

Institutions can also increase the incentives for reciprocal behavior, which, as game theorists have found, is a major factor toward cooperation. International institutions could be especially useful in making specific reciprocity more effective (as opposed to general reciprocity) by reducing transaction costs, limiting strategic options, and providing information to the players.[20] In fact, facilitating the exchange of information is institutions' most important function. Keohane and Martin argue that the main difference between neoliberals and neorealists is that the latter treat information as a constant, while the former look at it as a variable.[21] If institutionalists are right, then the level of information could be altered, and international organizations could be the vehicle for that.

Knowledge about others' capabilities and intentions is, indeed, a critical component of stable relationships. This was the case, for example, with the US-USSR arms race, and so the two countries established a hotline in order to improve the communication channels between the two governments. In 1962, during the Cuban missile crisis, information about the other side's intentions was very critical for the final outcome. If there were perfect knowledge about others' capabilities and intentions, conflict would be much less prevalent. On the eve of the First World War, for example, there were widespread misperceptions regarding other states' capabilities and intentions, and their hostility was highly exaggerated.[22] There are numerous examples of the utility of information in the security field, but it is also an important factor in the field of political economy. The daily functioning of global markets often creates risks and challenges for national economies, which could be mitigated through the regular exchange of information. This realization has prompted states to create several forums outside the extant IOs that are designed to exchange views and coordinate action, such as the G-7 and G-20.

Institutions in the Oil Issue-Area

As in all other areas of international life, information is also very important in the field of oil supply politics. Timely and accurate knowledge about variables like the level of oil reserves, consumption, trade and spare capacity is one of the factors that make oil markets function properly. At times of shortage, exchange of information becomes even more crucial. This is why states have attempted to coordinate policies within a number of bodies, like the International Energy Agency, and lately the International Energy Forum, which, for the first time, unites consumers and producers together.

In the cases examined in this study, states tried to coordinate policy in four different institutions. In 1951 and 1956, the OEEC oil committee was in charge. At the time, this was the only international oil body. The OEEC was succeeded by the OECD, and its oil committee, which were active during the 1967 and

1973 Arab embargoes. In the second case, the European Community also got involved. This was the first test for its newly adopted emergency system. The 1973 oil crisis had disastrous consequences for the world economy, leading to a prolonged recession. It was then decided that a full-fledged organization (as opposed to a committee) might be better suited to deal with oil market disturbances. In 1974, the International Energy Agency was created. Unfortunately, it has had ample opportunities to demonstrate its utility—in relation to the shortages provoked by the Iranian revolution, the Iran-Iraq war, the First Gulf War, and a number of other minor shortages not covered by this study.

If the institutionalist assumptions held, cooperation in the oil issue-area would have grown stronger in the 1970s and 1980s. The OEEC oil committee was an advisory body, compiling statistics and publishing reports. Its most important task was to coordinate the expansion of European refinery capacity, which it accomplished quite successfully. The Committee had no prerogatives, however, over oil crisis management. There was not an emergency procedure in place, or any provisions for coordination of oil policies. Its successor—the OECD oil committee—had a mixed record. It had a loose emergency system in place, which states tried to reform in 1972, without success. The IEA had a more elaborate emergency procedure, clearly defining what constitutes cooperative behavior in a crisis situation. It also provided higher transparency and increased cooperation in times of normal oil flows. All three institutions had permanent secretariats and established patterns of regular high-level meetings. Moreover, the shadow of the future was sufficiently long, as these countries had to interact on a long-term basis and in different issue-areas, being interlocked in a network of institutions. Most of the states, affected by the oil crises were also members of the European Community, which had an even more institutionalized structure, and which provided plenty of opportunities for issue-linkage. Even without institutional membership, these states already had intensive and repeated interactions, having lived next to each other for centuries and being each other's biggest trade partners. According to institutionalist theory, all these factors constitute a very rich ground for cooperation. Thus, institutionalists would predict that some form of cooperation would emerge in times of energy shortage.

The Security versus Economy Debate

An oil supply crisis, however, could severely damage a country's economy. Even the state's own survival could be in danger, oil being the lifeline of an industrialized economy. Insecurity about the future is so big and the stakes are so high that other considerations remain secondary. This characteristic of oil, neorealists would argue, alters the dynamics of cooperation. More specifically, it changes the payoff structure, so that the reward for cooperation (sharing oil and eventually running out of it) is much smaller compared with the punishment for defection (keeping one's oil reserves and accumulating more). Indeed, part of the debate on inter-state cooperation has focused on whether cooperation patterns differ in the international political economy and security realms. As Jervis asserts, some of the different conclusions of neorealism and neoliberalism come from their focusing on different issues, respectively, security vs. political economy or distribution vs. efficiency.[23] In addition, neoliberals have studied more areas, in which cheating does not threaten states' survival.[24] Mearsheimer even declared that institutionalism "ignores security issues" because of their conflictual character.[25]

On the other side of the debate, Axelrod and Keohane note that institutionalist theory can be applied equally to security and political economy issues.[26] In "The Promise of Institutionalist Theory", Keohane and Martin assert that such a dividing line does not exist, because of institutionalism's emphasis on information, which is a key variable in both areas.[27] The idea that cooperative behavior is more likely to arise in non-security areas, however, seems to follow from one of the institutionalist assumptions—that only mixed motives games are conducive to cooperation. These types of situations involve a certain degree of agreement prior to attempts at cooperation. Security issues are considered to be conflictual in essence, involving a higher level of competitiveness and much higher stakes than economic issues.[28] Indeed, if a state complies with a disarmament regime, while its partners defect, it risks its existence. This does not explain, however, the relatively similar levels of defection in security and economic regimes. For example, the United States has, on several occasions, refused to comply with WTO rulings. Similarly, Iran is trying to develop nuclear weapons, despite its NPT membership. What it explains is the lower prevalence of security institutions, in comparison with institutions in the field of economics.

The supply of oil stands at the intersection of the areas of political economy and security. Oil is a free-traded commodity subject to the laws of supply and demand. Oil prices and deliveries are determined on major stock exchanges, and on the basis of economic variables like quantities supplied and demanded. Regular oil deliveries, however, are essential for states' survival and could be seen as part of the "vital interests" of nation-states.[29] In the language of Gilpin, energy combines two important features—the pursuit of security and of welfare. The achievement of both is highly dependent on the availability of energy supplies.[30] Energy supplies are, thus, one of the few factors, which are critical for states' security, but at the same time are subject to the vagaries of the market. This makes energy and its supply a very sensitive issue for nation-states, where Pareto-optimal solutions are difficult to achieve without coordination.

In an anarchic environment, states need to fend for themselves, and a simple miscalculation or mistake could be costly. Even institutions, with all the tools they provide to enhance cooperation, cannot alleviate states' fear for their well-being and survival. Moreover, crises situations usually create powerful incentives for defection, since cooperation becomes riskier in the short run and less valued over the long haul, while gains from defection are available immediately.[31] In game theory terms, this situation resembles games like prisoner's dilemma or stag hunt, where actors have incentives to cooperate only if they are certain that everybody else will do so.

The Role of the Leader

Neorealists are quite skeptical about the opportunities for cooperation in interstate relations and the role of international institutions. Some of them, however, have suggested one factor that could help foster both—the leadership of a strong power. International organizations can only be effective as long as they are supported by "the principal states."[32]

This argument follows from hegemonic stability theory, which was first advanced by Charles Kindleberger in his seminal analysis of the Great Depression of the 1930s.[33] Kindleberger explained the economic crisis with the lack of a strong leader.[34] Britain was no longer able to lead, while the United States was not yet willing to do so. The author summarizes the important elements of hegemony as follows: the maintenance of a relatively open market, and the provision of long-term lending and discounting in crises. In other words, great powers maintain order by assuming a disproportionate share of the burden of maintaining it. They have the resources to establish and enforce the basic rules of governance. They also have the capabilities to discipline others and maintain compliance with these rules, in the face of coordination problems, such as cheating, free riding, and relative gains concerns. Hence, leaders can coerce others to cooperate according to the rules that they have established. If sharing oil supplies in the event of an emergency is one of those rules, it could only be enforced under strong leadership. In short, the international system is less anarchic and more stable under hegemony.

Scholars have found some evidence that a hegemonic system is more stable.[35] The majority of the studies focus on Great Britain in the nineteenth century and the United States after the Second World War. Several authors extend their analyses to the Netherlands in the seventeenth century, as well.[36] Most of these studies have found that a unipolar system is more likely to lead to the establishment and maintenance of free trade practices. Stephen Krasner, for example, found that an open trade system is most likely to be achieved when "a hegemonic state" is in its ascendancy.[37] The strong power can use positive and negative incentives, as well as its large economy and strong currency, to support free trade. It has the willingness to establish and maintain rules, not because it is altruistic, but because it is in its interest to do so, and it will continue to do that until the costs outweigh the benefits.[38] The United States' withdrawal from the Bretton Woods exchange rate system is a good illustration of this point. The system, created in the years after the Second World War, established the United States dollar as the world reserve currency, in which most international transactions were made. To maintain the system, the United States had to provide liquidity, and consequently run a balance of payments deficit, which was not sustainable in the long run. As a consequence, in 1971, Nixon unilaterally withdrew.

With respect to oil crises, neorealists would expect more cooperation during the 1950s and 1960s, as the United States was the undisputed leader of the issue-area at the time. The United States would enforce cooperative behavior to safeguard its main security interest—the protection of its territorial integrity. The unity of the Western bloc was an essential component of this defense. American leaders have, on several occasions, stated that if the Western world lost its oil supplies for a prolonged time, it would become vulnerable to attack. In the 1970s and the 1980s, while still being the leader of the Western bloc, the United States no longer had the oil resources necessary to enforce agreements. According to Yergin, until the 1970s, the "postwar petroleum order" was sustained by "an American-British ascendancy."[39]

Neoliberals have agreed that the dominant power plays an important role in facilitating the establishment of institutions but have counterargued that, because institutions are easier to maintain than to create, they can continue functioning even after the end of hegemony.[40] Leaders do play a role in fostering cooperation and institution-building, but once in place, the process can be sustained even after their decline. This is the case, for example, with the postwar trade regime.[41] US leadership was crucial to the creation of that regime, but not so crucial to its maintenance; hence the regime's continued existence, even after the decline of US power. With respect to institutions, it is less costly for states to continue co-operating within the already established ones than to recalculate their interests and establish new ones. Communication networks and rules are already in place, and the costs of maintenance are lower, while the risks and costs of a regime breakdown are very high.

Hegemonic stability theory has been widely contested. While there is, indeed, some evidence of the link between hegemony and free trade, it has been more difficult to establish such a link between hegemony and cooperation in other issue-areas or between hegemony and institutions. The latter is especially difficult, since there is only one case to study—the United States in the twentieth century. Although, sometimes it is possible to draw conclusions from a single case study, it is difficult to account for idiosyncrasies. Ruggie, for example, argues that the liberal economic postwar order was a result of the hegemon's identity—the Unites States' values and ideas, and that it would have been different if another power was at the helm.[42]

Hegemonic stability theory has received a number of criticisms, even as it applies to free trade practices. Some authors have observed that periods of hegemony do not coincide with periods of open trade systems, others have found that a relatively greater number of powers in the system is, actually, more conducive to free trade.[43] Recently, the most controversial question has been whether American power has declined in the last several decades. After an initial wave of scholarship announcing the end of American hegemony, a second wave followed, pronouncing the United States the most powerful actor since the Roman Empire.[44] This debate is a direct consequence of the ambiguity of the term hegemony and its difficult operationalization and measurement. The concept of power, itself, is difficult to measure, but it is even more difficult to assess what level of power superiority over others is necessary in order to control outcomes.

In an attempt to avoid these ambiguities, this study uses the more concrete notion of power in a particular issue-area. It is a more straightforward concept, easier to define and measure. A hegemon is traditionally defined as a state that has control over raw materials and markets, sources of capital, and a competitive advantage in the production of highly valued goods.[45] Such a state would be able to extend its influence on a global scale, and across issue-areas. It is also possible, however, that a state is disproportionately powerful in one issue-area but less powerful in others. In fact, this is more often the case in international politics. For the purposes of this study, a hegemon is defined as the state having control over resources in the oil issue-area. Such an area-specific version of hegemony is more suitable for the present analysis and has already been discussed by other authors. There is a general agreement that state power varies across issue-areas.[46] In addition, some scholars maintain that the realist assumption of fungibility should be relaxed since most states' resources vary from one issue-area to another and transferring power "from one area of activity to affect outcomes in other areas is difficult."[47] Thus, Germany has more power in Euro politics, and less in the field of nuclear weapons. Similarly, Saudi Arabia is more powerful in the oil than in any other field. Some scholars have already looked at the effects of polarity within issue-areas. Keohane, for example, has studied the changes in the regimes for oil, monetary affairs and trade in goods in the 1970s produced as a result of the decline of American preponderance in these three areas.[48]

It might be challenging to see how this discussion applies to US power in the post–World War II period. In the 1950s and 1960s, the United States harnessed power across the board—in oil, finances, arms, development, and almost every other field. It was the undisputed world leader. The United States might have linked different issues to coerce its partners to cooperate, using its power leverage in a particular issue-area or as a whole. In some cases, it might also be difficult to distinguish between power resources in an issue-area and power considered as a whole. However, from government documents, Congressional hearings and other resources, it becomes clear that the United States used its leverage in the oil issue-area to coerce European consumers. Measures such as setting up oil emergency committees, coordinating with the oil industry, granting the latter an antitrust exemption, negotiating with the Texas Railroad Commission on production increases and finally, shipping these increases to Europe show the use of oil leverage.

The American power in oil was mainly the result of the interplay of two factors. First, until the mid-1950s, the United States had the highest oil reserves, an almost negligible difference between domestic production and consumption, and a high domestic spare production capacity. The latter ensured that in the event of an emergency, production could be quickly increased to cover both domestic and foreign needs. In 1950, the United States provided 52% of the world's crude oil production, in 1997, just 10%.[49] Until 1963, surplus capacity averaged 4 million barrels per day. In March 1971, Texas was producing at 100% capacity for the first time.[50] In 1970, oil production reached its peak of 11.3 million barrels per day, and started declining afterward, while consumption kept moving upward. In addition, in 1973 imports already constituted 35% of oil requirements, while in 1967 they were only 19%. In 1980, the United States consumed 17 million barrels per day, while producing only 10 million. Unlike the 1950s and 1960s, the United States could no longer serve as a market balancer and supplier of last resort. Table 2, on the next page, provides a historical overview of United States production, consumption and reserves.

Table 2. United States production, consumption and reserves

Year	Production (million barrels per day)	Consumption (million barrels per day)	Production (% of world production)	Imports (% of consumption)	Proved reserves (million barrels)
1950	5.9	6.5	52	13	25,268
1955	7.6	8.5		15	30,012
1960	8	9.8	38	19	31,613
1965	9	11.5	30	21	31,352
1970	11.3	14.7	24.6	33	39,001
1975	10	16.3	18.9	37	32,682
1980	10	17	17	41	29,805
1985	10.1	15.7	19.6	32	28,416
1990	8.9	17	14.7	47	26,254
1995	8.3	17.7	13.3	50	22,351
2000	7.7	19.7	11.3	58	22,045
2005	6.9	20.8	9.4	66	21,757

Source: Data from U.S. Energy Information Administration, "U.S. Field Production of Crude Oil," http://tonto.eia.doe.gov/dnav/pet/hist/LeafHandler.ashx?n=PET&s=MCRFPUS1&f=M (accessed 09 March 2011), U.S. Energy Information Administration, "U.S. Product Supplied of Crude Oil and Petroleum Products" http://tonto.eia.doe.gov/dnav/pet/hist/LeafHandler.ashx?n=PET&s=MTTUPUS2&f=M (accessed 13 February 2011), U.S. Energy Information Administration, "Short-term Energy Outlook: February 10, 2010 Release" http://www.eia.doe.gov/pub/forecasting/steo/oldsteos/feb10.pdf (accessed 13 February 2011)

In November 1968, American representatives warned their European partners in OECD that American production would soon reach full capacity and that the security cushion they had been enjoying would be lost.[51] A government report published in 1970 stated that in the event of an embargo, the United States would be incapable of supplying the Eastern hemisphere with its oil requirements, as it had done before. The report advised Western European countries to develop their own emergency sources.[52]

The second factor was government cooperation with American oil corporations. Five of the "Seven Sisters" were American. While these companies were not government-owned, they did cooperate closely with the Administration, and in many cases, especially in times of crisis, advanced government interests. Louis Turner argues that, the United States, Britain and the Netherlands, played an important role in the oil industry until the 1970s, through their "parenthood" of oil majors.[53] On two occasions—1951 and 1956—companies got an exemption from antitrust legislation—a very critical move with regard to the oil crisis. In fact, government-business cooperation proved to be one of the most important factors during the Suez crisis. In the late 1960s and early 1970s, however, the balance of power shifted from the oil corporations to the producing countries. A wave of nationalizations took place, and producer governments took control over the oil fields and production. Libya, Algeria, Venezuela, Iraq and Kuwait, one after the other, announced nationalization programs in the course of the early 1970s. The creation of OPEC in 1960 contributed to this shift, as producer states were now able to exert more influence on the market by coordinating their policies. The United States lost a very important source of leverage.

The United States' lead in oil in the 1950s and 1960s is undisputable. During this time, when only two other countries had powerful oil companies capable of mastering resources on their own—the UK and the Netherlands (BP and Shell, respectively)—the United States stood far ahead of the other oil producers in terms of production and spare capacity. Table 3 provides a summary of world production and reserves in 1960.

Table 3. 1960 production and reserves—selected countries

	Production (million barrels per day)	Reserves (billions of barrels)
Canada	0.52	10.2
Iran	1.07	65
Iraq	0.97	29
Kuwait	1.69	64
Mexico	0.27	2.8
Nigeria	0.02	15
Russia (former USSR)	2.91	75
Saudi Arabia	1.31	138
United Arab Emirates	0	20
United Kingdom	0	-
United States	8	36.8
Venezuela	2.85	13.7

Source: Data from U.S. Energy Information Administration, "Energy Perspectives," http://www.eia.doe.gov/emeu/aer/txt/ptb1105.html (date accessed 15 November 2010)

Conclusion

This study compares neorealism and neoliberalism, by applying some of their assumptions to the issue-area of oil. More specifically, I examine under what conditions states cooperate. The two theories offer contending explanations. Neorealists argue that states would only cooperate to further their own interests, or when coerced by a dominant power. Similarly, international institutions are only effective when supported by the strong states of the day. Neoliberal institutionalists, on the other hand, believe that international organizations can foster cooperation through a variety of mechanisms, including exchange of information, iterated interaction and issue linkage. I apply these assumptions to seven cases of oil supply disruptions, looking at the role of four different institutions and the United Stated as the leading power in oil until 1970. This allows me to test these theories in a structured and controlled way.

If neorealists were correct, I would expect to find more cooperation during the pre-1970 cases, when the United States was at the height of its power. If neoliberals were right, I would expect cooperation to be more prevalent in later crises, as institutional mechanisms grew stronger and the European Community adopted its oil emergency policy.

Undoubtedly, international organizations do play a role in international politics. They provide a setting for the exchange of information, reduce transaction costs, and, in rare cases, could even alter states' understanding of their interests. However, there is high variation across institutions and cases (within the same institution), in terms of their effectiveness in bringing about cooperative behavior. This is why it is interesting to study under what conditions institutions foster cooperation. This overarching puzzle is easily transferable to the issue-area of oil. Why is it that in 1973 oil consumers disagreed, while in 1967 they peacefully coordinated their efforts? In both cases the OECD oil committee was in charge. This question has important implications, given the tremendous impact an oil supply disruption could have on the economy. The next four chapters will try to answer some of these theoretical puzzles by looking at seven post–World War II supply disruptions. I first turn to the Suez crisis of 1956.

Notes

1. Stinchcombe, *Constructing Social Theories*, 25.
2. Robert O. Keohane, *After Hegemony: Cooperation and Discord in the World Political Economy* (Princeton, NJ: Princeton University Press, 1984), 51.
3. Oye, *Cooperation Under Anarchy*, 5.
4. Keohane, *After Hegemony*.
5. Andrew Moravcsik, "Taking Preferences Seriously: A Liberal Theory of International Politics," *International Organization* 51, no. 4 (1997); John Gerard Ruggie, "What Makes the World Hang Together? Neo-Utilitarianism and the Social Constructivist Challenge," *International Organization* 52, no. 4 (1998).
6. Robert Jervis, "Realism, Neoliberalism, and Cooperation: Understanding the Debate," in *Progress in International Relations Theory: Appraising the Field*, ed. Colin Elman and Miriam Fendius Elman (Cambridge, MA: MIT Press, 2003), 286.
7. Joseph M. Grieco, "Anarchy and the Limits of Cooperation: A Realist Critique of the Newest Liberal Institutionalism," *International Organization* 42, no. 3 (1988); Kenneth N. Waltz, *Theory of International Politics* (Reading, MA: Addison-Wesley Pub. Co., 1979).
8. Keohane, *After Hegemony*.
9. Robert M. Axelrod, *The Evolution of Cooperation* (New York: Basic Books, 1984).
10. Oye, *Cooperation Under Anarchy*.
11. Charles Lipson, "International Cooperation in Economic and Security Affairs," *World Politics*, 1984.
12. Ibid., 19.

13. Joseph M. Grieco, *Cooperation Among Nations: Europe, America, and Non-tariff Barriers to Trade* (Ithaca, NY: Cornell University Press, 1990), 10.

14. Ibid., 11.

15. Inis L. Claude, *Swords into Plowshares; The Problems and Progress of International Organization* (New York: Random House, 1971); Grieco, "Anarchy and the Limits of Cooperation"; Stephen D. Krasner, "Global Communications and National Power: Life on the Pareto Frontier," *World Politics*, 1991; John J. Mearsheimer, "The False Promise of International Institutions," *International Security* 19, no. 3 (1994); John J. Mearsheimer, *The Tragedy of Great Power Politics* (New York: W.W. Norton, 2003); Kenneth N. Waltz, "Structural Realism after the Cold War," *International Security* 25, no. 1 (2000); Waltz, *Theory of International Politics.*

16. Keohane and Martin, "Institutional Theory as a Research Program"; Keohane, *After Hegemony*; Robert O. Keohane, *International Institutions and State Power: Essays in International Relations Theory* (Boulder, CO: Westview Press, 1989); Robert O. Keohane, Joseph S. Nye, and Stanley Hoffmann, *After the Cold War: International Institutions and State Strategies in Europe, 1989–1991* (Cambridge, MA: Harvard University Press, 1993); Helen V. Milner, "Power, Interdependence, and Nonstate Actors in World Politics: Research Frontiers," in *Power, Interdependence, and Nonstate Actors in World Politics*, ed. Helen V. Milner and Andrew Moravcsik (Princeton, NJ: Princeton University Press, 2009).

17. Keohane, *After Hegemony.*

18. Ernst B. Haas, *The Uniting of Europe; Political, Social, and Economic Forces, 1950–1957*, The Library of World Affairs, no. 42 (Stanford, CA: Stanford University Press, 1958).

19. Oye, *Cooperation Under Anarchy*, 12–17.

20. Robert O. Keohane, "Reciprocity in International Relations," *International Organization* 40, no. 1 (1986): 25.

21. Keohane and Martin, "Institutional Theory as a Research Program," 79–80.

22. Stephen van Evera, "Why Cooperation Failed in 1914," *World Politics* 38, no. 1 (1985).

23. Jervis, "Realism, Neoliberalism, and Cooperation," 283–4.

24. Ibid., 290.

25. Mearsheimer, "The False Promise of International Institutions," 15.

26. Keohane, *After Hegemony*, 227.

27. Keohane and Martin, "The Promise of Institutionalist Theory."

28. Robert Jervis, "Security Regimes," *International Organization* 36, no. 2 (1982).

29. Robert Gilpin, *War and Change in World Politics* (Cambridge, MA: Cambridge University Press, 1981), 25.

30. Gilpin argues that states do not necessarily pursue only security, but a combination of security and domestic welfare, depending on the perceptions of the ruling elites and the international environment.

31. Lipson, "International Cooperation in Economic and Security Affairs," 22.

32. Waltz, *Theory of International Politics*, 88.

33. The term hegemony was first used by Marxist thinkers, for example Antonio Gramsci, who developed the concept of cultural hegemony (see Gramsci 1992). In this volume, it is used in the context of hegemonic stability theory.

34. Charles P. Kindleberger, *The World in Depression: 1929–1939* (London: Penguin Press, 1973), 292–306.

35. Gilpin, *War and Change in World Politics*; Scott C. James and David A. Lake, "The Second Face of Hegemony: Britain's Repeal of the Corn Laws and the American Walker Tariff of 1846," *International Organization* 43, no. 01 (1989).

36. Torbjorn L. Knutsen, *The Rise and Fall of World Orders* (Manchester: Manchester University Press, 1999); Peter J. Taylor, *The Way the Modern World Works: World Hegemony to World Impasse* (Chichester: Wiley, 1996).

37. Krasner, "Global Communications and National Power: Life on the Pareto Frontier," 323.

38. Gilpin, *War and Change in World Politics.*

39. Daniel Yergin, *The Prize: The Epic Quest for Oil, Money, and Power* (New York: Simon & Schuster, 1991), 565.

40. Keohane, *After Hegemony*, 50.

41. Charles Lipson, "The Transformation of Trade: The Sources and Effects of Regime Change," *International Organization* 36, no. 2 (1982).

42. Ruggie, "What Makes the World Hang Together?," 863.

43. See, for example, Edward D. Mansfield, "The Concentration of Capabilities and International Trade," *International Organization* 46, no. 3 (1992).

44. Andrew Hurrell, "Hegemony, Liberalism and Global Order: What Space for Would-be Great Powers?," *International Affairs* 82, no. 1 (2006); G. John Ikenberry, *America Unrivaled: The Future of the Balance of Power*, Cornell Studies in Security Affairs (Ithaca, NY: Cornell University Press, 2002); Susan Strange, *The Retreat of the State: The Diffusion of Power in the World Economy* (New York: Cambridge University Press, 1996); Waltz, "Structural Realism after the Cold War."

45. Keohane, *After Hegemony*, 32.

46. David A. Baldwin, "Power Analysis and World Politics: New Trends versus Old Tendencies," *World Politics* 31, no. 2 (1979); Robert O. Keohane and Joseph S. Nye, *Power and Interdependence* (New York: Longman, 2001); Milner, "Power, Interdependence, and Nonstate Actors in World Politics."

47. Robert O. Keohane, "Theory of World Politics: Structural Realism and Beyond," in *Neorealism and Its Critics*, ed. Robert Keohane (New York: Columbia University Press, 1986), 187.

48. Keohane, *After Hegemony.*

49. U.S. Energy Information Administration, "Major Disruptions of World Oil Supply," http://www.eia.doe.gov/emeu/25opec/sld001.htm (accessed 10 July 2009).

50. Yergin, *The Prize*, 567.

51. Ibid., 568.

52. Kapstein, *The Insecure Alliance*, 156–7.

53. Louis Turner, "The Oil Majors in World Politics," *International Affairs (Royal Institute of International Affairs)* 52, no. 3 (1976): 370.

Chapter 3
The Suez Crisis

The Suez Crisis was one of those defining events in history, which keep resurging in political and academic debate. The 1956 Anglo-French debacle has a very important place in the history of the Western world, the Cold War and trans-Atlantic relations. For Britain and France, it was a moment of realization of the new power realities in the international system. For the United States, it posed the beginning of a new stage in American involvement in the Middle East, the so-called Eisenhower Doctrine: containing Arab nationalism and limiting Soviet influence in the region. From a Cold War perspective, the crisis unfolded into an East-West confrontation, with the two sides coming close to open hostilities.

In terms of energy security, the Suez events led to the biggest oil shortage for European countries at the time. The closure of the Canal was followed by a Saudi embargo to France and the UK, and a sabotage to two pipelines of the Iraq Petroleum Company (IPC) delivering oil to Mediterranean ports. This brought to a halt all flows of Middle Eastern oil to OEEC Europe, amounting to about 2 mbd or close to 90% of regular deliveries.[1] In relative terms, this was the biggest oil shortage in the postwar period. Yet, it had a much smaller impact on the consumer's world, compared to later crises. Undoubtedly, the relatively low reliance on oil in Western Europe was an important factor. In 1955, Europe was satisfying only 18% of its energy needs with oil. By 1970, this number had increased to 60%.[2] Nevertheless, it is still remarkable that consumers did not scramble for supplies, and avoided competitive bidding, given the magnitude of the shortage. After all, they did it later, in 1978 and 1980, when the relative shortage was smaller.

Unlike in the 1970s, this time, all factors necessary to assuage market participants' fears were present. First, there was a largely accepted expectation that an oil shortage would be overcome by increased production in the Western hemisphere, and in particular the United States.[3] This provided a highly needed psychological buffer. At the time, production in Texas could be increased in a relatively short time, which eventually the Texas Railroad Commission did in late January, after considerable government pressure. Second, the US administration, in consultation with the oil majors started preparing for emergency early on. An Action Plan was on the table on 10 August 1956 and the Middle East Emergency Committee had its first meeting on 24 August—two months before the actual closure of the Canal. This sent a positive signal to allies across the Atlantic, providing them with a sort of insurance policy. It also gave them the time to discuss emergency schemes within the OEEC oil committee, which took a very active role in the course of the emergency by taking part in oil reallocation. This was the only time an intergovernmental body, in consultation with industry, participated in oil redistribution. Last but not least, was the Administration's insistence to interact with European governments en bloc and not on a country-by-country basis. As America was the supplier of last resort at the time, there was no other option but to follow.

The role of the oil companies in 1956, as in most other oil shortages, is indisputable. What distinguished this case from later ones is the close collaboration of the majors with the American administration, and the fact that they were still in control of the entire chain of oil production and distribution—an advantage they lost in the early 1970s. The antitrust exemption they were granted early on allowed them to share information among each other and with their European counterparts, which was essential for overcoming the crisis.

Crisis Background

Events unfolded very quickly between July and December 1956, but the crisis has been brewing for months before that. Openly anti-Western and increasingly nationalistic, with ambitions for pan-Arab leadership, Egyptian president Gamal Abdel Nasser signed an arms deal with the Soviets, and opened diplomatic relations with China. Nasser was one of the leaders of the Egyptian revolution in 1952, which overthrew King Farouk and set the stage for reforms, including a ban on political parties. He later on became one of the central figures of the Arab world, advocating Arab nationalism and anti-imperialism.

On July 26, 1956, President Nasser unilaterally declared the nationalization of the Suez Canal, previously owned by the Anglo-French Suez Canal Company. Britain was the largest shareholder in the company. According to an 1888 international convention, the Canal was to be free and open to every ship regardless of nationality.[4] In his speech, announcing the nationalization and later on, in correspondence, President Nasser declared that he would "safeguard the freedom of passage" through the canal and would compensate the previous shareholders.[5]

The nationalization of the Canal was the immediate reason for the joint French-British-Israeli attack on Egypt at the end of October. The Egyptian authorities immediately blocked the Canal by sinking ships and wrecks along its entire length. France and Britain landed in Egypt on November 5, after several days of aerial assault. After significant pressure from the international community and, in particular, the United States, a cease-fire was signed on November 6, and France and the UK withdrew by December 25. Israeli troops followed in March 1957. Two months later, the Canal was open for shipping again.

The Suez Canal had an indisputable economic value at the time. It provided the main trade link between Europe and Asia, and a major oil route. Suez was not the only concern, however. The reasons for the joint attack were more complex and each of the three countries had its own underlying motives. France, at the time, was entangled in the Algerian war for independence, which President Nasser supported.[6] The French government was suspecting that Nasser was training Algerian military leaders and rebels. An attack on Egypt, it was thought, would strengthen French positions on the Algerian front. In the words of Irwin Wall, "Suez had always been about Algeria."[7] In the same vein, an editorial in *Le Monde* argued that "Paris was not so much interested in Suez but in the need to make Nasser lose face."[8] Israel had been preparing an assault on Egypt for several months already. The government believed that Nasser was planning to attack Israel long before the Suez crisis.[9] For Great Britain, ownership of the Canal was a matter of both economic concerns and prestige. It was one of the main trade routes for Asian goods, but it was also part of the British-influenced Middle East. Losing the Canal meant discrediting the British and their position as a leading power in the region. Some authors also maintain that securing oil supplies was the main factor behind the British decision to attack, and oil remained a primary concern for the government throughout the crisis.[10]

The Suez crisis qualifies as one of the most critical junctures of the Cold War. One of its most important consequences was the realization that middle powers could no longer act alone, but only in concert and with the consent of the Great Powers. Even more important, however, was the awakening of the UK and France to the fact that they no longer belonged to the club of the strongest. This had implications for future political developments in both countries. Egypt and, in particular, its leader Nasser emerged as the winners, and the president asserted his place as one of the leaders of the Arab world. Last but not least, Suez saw the beginning of UN peace-keeping with the establishment of the UN Emergency Force in November 1956 to monitor the peace between Israel and Egypt.

The Oil Problem

The Suez Canal at the time was a critical link in the transport of oil to Europe. At the moment of the crisis, OEEC Europe received oil from several locations, including Iraq, Saudi Arabia, and North and South America. The Canal and two pipelines were the main routes of transportation for Middle Eastern oil. One of the pipelines was sabotaged on Syrian territory shortly after the closure of the Canal. Thus, of all oil routes only the Tapline, passing through Saudi Arabia, Jordan, Syria and Lebanon, was still in service. It was, however, closed for France and the UK, due to the Saudi embargo. Of the 2.3 million barrels per day Europe consumed in July 1956, 1.3 million passed through Suez and about

700,000 were transported through the above pipelines. This represented a 90% reduction of Europe's overall oil supplies, although the impact varied from country to country.

Naturally, France and the UK suffered more than the others as they were direct participants in the conflict and did not receive oil even through the non-interrupted channels. Different import, consumption and energy production patterns also played a role. For example, France received almost half of its oil supplies through Suez, the UK—two thirds, while Germany—only 20%.[11] According to an OEEC report, in 1954, the UK produced 0.06 million tons of petroleum, Germany—2.7 and France—0.6.[12] Germany, at the time, was the second biggest oil producer in Europe, after Austria, whereby domestic production satisfied a third of local demand. Moreover, petroleum accounted for only 9% of the energy mix there, while it was 22% in France, and 13% in the UK (see table 4).[13] All these conditions made France more sensitive to the disruption than the other European consumers.

Germany, on the other hand, was the least impacted. On 8 November 1956, *Frankfurter Allgemeine Zeitung* reported that tankers belonging to the German subsidiary of Esso were loaded at Sidon, in Lebanon, while French and British tankers were refused oil. According to an OEEC report, two other important factors helped to ease the shortage in Germany. The Federal Republic benefited first, from its own benzol production, and second, from significantly increased imports of some refined products like gasoline and diesel, from the "Soviet-occupied zone." At their 8th meeting, on December 17, the OEEC Middle East Emergency Committee members reported that West Germany was receiving almost 100% of its oil requirements, while other countries were getting very little.[14] This would also explain the relative lack of literature on the effect of the Suez crisis on Germany and its economy.

Table 4. France, Germany and the UK—selected oil data, 1956

	Oil in the energy mix (%)	Oil received through Suez (%)
France	22	48
Germany	9	20
United Kingdom	13	60

Source: Data from Organization for European Economic Cooperation, *Europe's Growing Needs of Energy: How Can They Be Met?* (Paris: OEEC, 1956), Organization for European Economic Cooperation, *Report by the Oil Committee to be Published in 1956* (Paris: OEEC, 1956)

The United States

The US government, in consultation with the American oil companies, took an active role both in supplying European countries with crude and products, and with solving the crisis overall. It was mainly under US insistence that the French and British troops left Port Said in December 1956. In fact, the American decision not to supply oil to Europe unless all troops withdrew was the main factor behind the end of hostilities. Eisenhower had clearly stated that the oil emergency plan would not be activated and no oil would be shipped to Europe unless the belligerents withdrew.[15] This was one of the first times in history that oil was used as a foreign policy weapon.

The American majors, in consultations with the three biggest European companies were able, in a remarkably short time to achieve three goals: increase tanker capacity, reroute oil destined to North America toward Europe, and increase Texas production. The first measure was of special significance, as oil was still available, but it had to be transported around the Cape of Good Hope, which increased the journey considerably. Overall, the Oil Lift, as it was later called, was a major operation, requiring the highest level of coordination between companies and governments. And while companies deserve to receive full credit for the handling of the crisis, this could not have been possible without the support of the American government, which granted an antitrust exemption to the oil majors, relaxed the requirement for US-flagged tankers, allowed a lower loadline for tankers, and pressured the Texas Railroad Commission into increasing oil production. All these moves proved indispensable in order to bring the Oil Lift to success.

Similarly to the Abadan crisis of 1951, the US government was concerned with a possible Soviet intervention. Most students of Suez agree that this was the main reason behind the US involvement in the crisis.[16] In any case, the American government was firmly opposed to the Franco-British-Israeli attack, and put all its diplomatic resources into persuading and coercing these countries into withdrawal. The American stance is very well summarized in a statement by Secretary of State Dulles of July 31 that a British decision to attack would be "out of date," would jeopardize the supply of oil, and would turn the world against the Western bloc.[17] The latter was of particular concern to the administration. It was feared that attacking Egypt, while at the same time, condemning the Soviet intervention in Hungary, would have devastating consequences for the authority and influence of the Western bloc. US concerns were, to some degree, validated by a letter sent by the Soviet premier Bulganin to the governments of France and the UK, threatening unilateral action. A separate letter was sent to President Eisenhower proposing a joint US-USSR intervention in Egypt. For Eisenhower, this was a proof of the Soviets' intentions to interfere in both Western Europe and the Middle East. The United States differed from Britain and France on one more point—the government did not consider the nationalization of the Canal an illegal action, as long as it continued its normal operations and the 1888 Convention remained in force.[18]

Aware of the importance of the Suez Canal as a petroleum transportation route, the American presidency was especially concerned with an eventual shortage of oil in Europe. The administration believed that an interruption of Middle Eastern oil supplies would be catastrophic for NATO security.[19] Indeed, at a time of fast economic recovery in Europe, with an increasingly threatening Soviet presence next door, a continuous oil interruption might have been devastating.

Despite this awareness, the swiftness of the US reply to the oil shortage in Europe is remarkable. A successful oil lift to Europe while keeping American demand satisfied required a reorganization of production, which, as Frankel has noted was "at short notice a formidable task."[20] The government, however, had the willingness and the resources to achieve that in a relatively short period of time.

Indeed, the administration organized rather quickly and there was an emergency Plan of Action at the table already in the beginning of August, three months before the halt of Suez shipping. The existence of an organizational structure in place, inherited from the Korean War, proved to be crucial in the timely handling of the crisis. Most important, however was the resolve of the American administration to help its allies. It was probably prompted by fears of Soviet intervention, but this is not the critical point here. The timely and swift American reaction helped avert a more serious oil crisis, and ensured cooperation on the other side of the Atlantic. The administration's insistence on assisting European governments en bloc and not individually proved to be crucial. An editorial in the *Frankfurter Allgemeine Zeitung* from November 7 called for the federal government to negotiate oil supplies with Washington separately, since a common European solution to the oil problem was not feasible at the moment.[21] At the end, it was through the coordination of several committees, on both sides of the Atlantic, that Europe could receive the highly needed oil.

In fact, the level of institutionalization of the emergency oil supplies to Europe was remarkable during Suez. Such a high level of coordination was not attained during any of the other postwar oil crises. There was a web of committees, set up both in the United States and Europe, which regularly exchanged information and participated in each other's deliberations. One very critical factor contributing to the success of this well-organized machinery was the prior existence of a domestic institutional structure in the United States.

The most important institution on the US side was the Middle East Emergency Committee (MEEC) set up shortly after the nationalization of the Canal, with the goal to "estimate the petroleum requirements of each foreign country or area to be supplied…and prepare and submit recommended schedules."[22] Its first meeting was held on August 24, 1956—more than two months before the actual start of hostilities and the closure of the Canal. Composed initially only of the 15 biggest oil companies with foreign operations, it was later reorganized to include some independents, due to antitrust concerns.

It is important to highlight the origins of the MEEC, which could be traced back to the Korean War. In order to streamline military resources for the war in Asia, in December 1950 President Truman established the Office of Defense Mobilization, which was supposed to coordinate all wartime mobilization activities, including industry and transportation. One of the most crucial among its responsibilities was the establishment of industry agreements in times of security crises. These agreements would allow the government to rely on industry resources, without "complete direction of the economy."[23] This provision proved to be very controversial later and was the subject of continuous intra-institutional struggle, especially between the Justice Department, and the Departments of State and Interior. Nevertheless, it played a critical role in the first two postwar oil crises—Abadan (1951) and Suez (1956). It allowed for the formation, in 1951, of the so-called Voluntary Agreement Relating to Foreign Petroleum Supply. This agreement gave the government access to crucial information regarding oil supplies. The Foreign Petroleum Supply Committee, which proved critical in alleviating the "shortages of petroleum supply in friendly foreign nations" during Abadan, was established in the framework of this agreement.[24] The latter was still in force in 1956, although considerably amended (in 1955), to allow for stricter government regulation. In order to successfully coordinate supplies, the Committee was granted immunity from US antitrust law—a very critical governmental move, which allowed companies to share information on production and supply. For example, companies exchanged regular information on their oil positions and the number and location of the tankers in their possession. In this way, the shortage of tankers—the biggest problem during Suez—could be overcome. To ensure the legality of all coordinated action, a representative of the antitrust division of the Justice Department was present at all MEEC meetings.

This process of consultation and coordination among companies and among companies and government executives was very successful at managing the European supply shortage. Overall, the prior existence of a strong institutional base in the United States and the granting of antitrust immunity to oil companies were important factors during the crisis. They allowed for prompt and efficient decision making. It is possible that their absence during later crises was part of the reason for the lower effectiveness of the US coordination efforts.

The Foreign Petroleum Supply Committee adopted its Suez Plan of Action as early as August 10, 1956.[25] The Middle East Emergency Committee was set up as part of the Plan. It had twenty meetings between August 1956 and May 1957. It had a permanent staff, and representatives from both the United States and European governments participated in its deliberations. The committee was suspended for a short time after the start of hostilities, but resumed in the beginning of December, after the Franco-British withdrawal declaration.

MEEC had the uneasy assignment of redistributing oil flows and channeling them toward the affected countries. This was a tremendous task, given the lack of reserve resources at the time and the fact that oil was supplied on the basis of long-term contracts, which could not be broken easily. Additional obstacles included the loss incurred by refineries from substituting Middle Eastern crude with other types of oil. But even harder was the task of finding additional tanker tonnage and availability. As Frankel would observe, the oil supply problem during Suez "was never one of a shortage of oil itself."[26] With the exception of the Saudi boycott to France and the UK, oil shipments were running, but they had to be rerouted around the Cape of Good Hope. This significantly increased the distance and additional tankers had to be put into operation, in order for the same amount of oil to be transported. One of the most important measures, adopted by the MEEC and discussed at large at every meeting, was tanker capacity. Rerouting of tankers and the use of American military–mothballed tankers considerably eased the burden of supplying Europe.

The Middle East Emergency Committee was also very successful in redistributing Middle Eastern oil bound for the United States toward Europe. At their 9th meeting on December 28, MEEC members reported that oil flows from the Persian Gulf to the United States have decreased from 127,000 b/d in the beginning of December to 85,000 b/d, with the difference going to Europe.[27] In March, oil moving toward Europe from all sources amounted to 2,203,000—the amount lost as a result of the Suez crisis.[28] The 1956 oil shortage had been successfully overcome. The role of the MEEC was acknowledged by the chairman of the OEEC oil committee who stated in April 1957 that the supply situation was going back to normal thanks to "the heavy additional imports from the Western hemisphere and the cessation of imports of Middle East oil into North America combined with considerable re-routing of tankers by means of cooperation between OPEG and MEEC."[29]

Besides supplying oil and tankers, and coordinating the emergency effort, the US government's role proved crucial in another aspect: assisting the UK government with the highly needed dollar reserves to buy oil. Initially, the administration refused to support the sterling, which experienced a significant downfall. In addition, the United States prevented a British withdrawal of funds from the IMF, where it had a predominant vote. According to Richardson, American financial pressure proved to be "one of the most powerful factors in securing a British surrender.[30]

On the other side of the Atlantic, coordination efforts developed slower, and it seems, mostly under US lead. There, as well, the pre-existence of some form of institutional base proved to be important. In the beginning of September 1956, an industry committee—the Oil Emergency London Advisory Committee (OELAC)—was set up in London. It included representatives of the three biggest European oil companies—Royal Dutch/Shell, British Petroleum and Companie Française des Pétroles, and foreign-based executives of the American multinationals. Its London location suggests that Washington preferred London as its base for communication and coordination with Europe. This was the case until the OEEC oil committee became more involved in the crisis. While OELAC was, initially, the only European body responsible for emergency supply, it got support by the British Oil Supply Advisory Committee (OSAC), which had been established during the Iranian crisis of 1951. OSAC had been set up as a domestic committee, since it was considered that Abadan was predominantly an internal issue. In 1956, however, it got involved in the international crisis effort.[31] After the OEEC Petroleum Emergency Group (OPEG) was formed in early December, OELAC was disbanded. OPEG had the same company participation, but was part of the OEEC oil committee, and not independent. This facilitated the coordination between industry and government.[32] The communication between MEEC on the American side and OPEG on the European continent proved to be crucial in the course of the Oil Lift. MEEC had a leading role in the operation. OPEG devised detailed plans about the needs of each European country, based on pre-Suez consumption, local production and stock levels.[33] The task of preparing allocation plans to distribute oil among the most needed fell on the OEEC oil committee.

The Organization for European Economic Cooperation

The OEEC represents the first attempt of European countries at institutionalized economic cooperation. Its creation was prompted by the need for a quick economic recovery after the Second World War. The US government insisted on a joint European recovery program and European unity was a condition for any American aid.[34] The organization was born out of the Conference on European Economic Cooperation in July 1947, which met to discuss the details of the US post–World War II aid to Europe. In 1948, when European governments were focused on national recovery, the OEEC was given only advisory functions. Not surprisingly, decisions were to be taken by unanimity.

The OEEC had very limited responsibilities in the oil area. The initial purpose of the organization was "the elaboration and execution of a joint recovery program," promotion of "the development of production.....and the maximum possible interchange of goods and services." The organization was also supposed to strengthen economic cooperation among its members, with the view of building a "free trade area." Its Oil Committee, which was the only international energy body at the time of the Abadan and Suez crises, was also formed to serve the needs of the disbursement of American aid. Set up in May 1948, its first task was to coordinate the distribution of funds earmarked for the establishment of a European refining industry. This was accomplished successfully with an increase of OEEC countries' refinery capacity from 19.5 million of tons a year in 1948 to 103 million tons in 1955. The Committee also compiled statistics on petroleum products and submitted recommendations to member-states on matters of oil use. Oil crisis management was not one of its functions. Ironically, Suez proved to be the highest point of an IO involvement in a crisis. The Committee elaborated emergency schemes and participated in oil re-allocation. Later, in the 1970s, while detailed emergency plans were available, such coordination never ensued. The success of the oil committee in 1956 could only be attributed to strong leadership in the face of the United States.

The Oil Committee

Unlike during the Abadan crisis five years earlier, in 1956, the OEEC oil committee got heavily involved in the organization of the crisis response. Its formal responsibilities had not changed between 1951 and 1956. It was understood, however, that the closure of the Canal could have much more severe implications for the European oil supply. There was also a higher awareness of the eventual consequences of a serious oil shortage. An oil committee report from October warned that "uncoordinated action to secure supplies might well damage the economic structure of Western Europe."[35] Consequently, shortly after the nationalization, the OEEC Council instructed the oil committee to prepare emergency allocation plans for Europe. Naturally, the committee started by collecting information on consumption and production, and for the first time—stock levels. Member-states were required to submit stock data every 10 days. Developing an emergency scheme, however, proved to be a laborious task, which took the committee several months. The first plans materialized in the beginning of January.[36] The reasons for this delay are manifold. In the first place, the intergovernmental structure of the committee, whereby national delegations had to consult with governments on each decision taken, and the latter were to be taken with unanimity, slowed down the process considerably. Second, disagreements of different sorts appeared in the course of the committee discussions. For example, countries could not agree on whether to communicate with the industry committees through the respective member-states (i.e., France, the Netherlands, the United States and the UK) or whether to directly link them to the oil committee deliberations.[37] This debate continued well into November, when it was decided to establish OPEG. In contrast, the American side had a schedule of diversion of oil from the Middle East to Europe in early December.[38] Increased supplies from US fields came later, due to the initial reluctance of the Texas Railroad Commission to increase production. This was done in late January, after considerable pressure by the administration.[39] Before that, companies were using stock oil to supply Europeans. And third, part of the reason for the delayed oil committee action rests with the fact that six different allocation methods were discussed, before a decision was taken. Apparently, it was difficult to reach a consensus on what would be the optimum allocation plan. According to the allocation scheme that won, shortages would spread equally among member-states with each country getting the same proportion of its consumption. Oil companies were in charge of distributing the available oil. In addition, 200,000 tons of oil was to be allocated directly by the oil committee, to the most "hardship cases," at the end of each planning cycle.[40] Cycles were 10 days long. This was the only time that the oil committee directly participated in oil allocations. In all other cases, oil companies, on their own, or through various committees, had managed the allocations. Undoubtedly, the involvement of this intergovernmental body, in 1956, was exceptional.

By March, Europeans were receiving almost the totality of their oil requirements. Both the pipelines and the Canal resumed activity in May and the various committees were dissolved. There was a suggestion of the British government to maintain OPEG on a standby basis, so that it is able to meet any future emergencies.[41] It was agreed among industry, however, that without the MEEC, OPEG could not operate efficiently.

The role of the oil committee should not be exaggerated. It did take part in the overall crisis management, elaborating an emergency scheme and directly participating in oil redistribution. This, by itself, is a feat to be recognized. In a time of severe oil shortage, which seriously threatened their economic survival, governments were able to negotiate oil allocations, sometimes renouncing their share to help a neighbor in need. This was cooperation at its highest. It defied anarchy, showing that more benign international relations were, indeed, possible. However, two other important variables need to be recognized here, as well. Cooperation within the oil committee was possible because of the security cushion, provided by the United States. The availability of spare production, and the shared expectations that the United States was willing and able to help, contributed to this insurance policy. In addition, as in all other cases, the oil committee reacted rather slowly. Consultations with governments at home were required at every step of the negotiations process. Had it not been for the insurance provided by the Americans, it is almost certain that these negotiations would have not taken place at all. The uncertainty of an oil crisis would have pushed rational governments to seek for oil unilaterally.

Conclusion

The 1956 Oil Lift was a remarkable achievement that was never repeated. It involved a tremendous effort of coordination among companies and governments, which was only possible due to the guidance of the American administration. As a result of the US Oil Lift, Europe was able to maintain a level of 90% of normal oil supplies throughout the crisis period. In short, "American interest backed by power was critical for the outcome of the crisis."[42]

The US role could be broken down to the following three elements: availability of spare production and tanker capacity, early-on emergency planning and a requirement to work with European governments en bloc. These three factors proved crucial in bringing cooperation, not only during Suez, but during all pre-1970 crises. The first two provided a security cushion for Europeans, which had the insurance that help was coming. The third gave them the opportunity to negotiate oil re-allocation within the OEEC oil committee.

The oil committee, albeit slightly late, got involved in the crisis management and participated in oil redistribution. This is the only time when an intergovernmental body was directly involved in reallocation. The emergency scheme was put in place by the oil companies, in close consultation with governments. This coordinated effort greatly alleviated the consequences of the crisis, and the price increase was minimal. At the height of the shortage, prices were about 40% higher, mostly due to increased freight costs. In comparison, in 1973, prices increased as much as 400%.

This case clearly lends support to the power paradigm. The American ability and willingness to serve as a supplier of last resort was essential in fostering cooperation among allies, both within and outside of the oil committee. It is very likely that without the American coercive leadership, Europeans would have competed for supplies, exacerbating the crisis. There were signs of that in the beginning when Britain demanded special treatment, due to its "special relationship," and Germany wanted to negotiate American oil supplies unilaterally. Fortunately, a major scramble was avoided, thanks to American leadership. Events developed somewhat differently in 1973—during the Yom Kippur Arab embargo. This critical case is the topic of the next chapter.

Notes

1. U.S. Congress. Senate, *Emergency Oil Lift Program and Related Oil Problems: Joint Hearings Before Subcommittees of the Committee on the Judiciary and Committee on Interior and Insular Affairs*, 85 Cong., 1 sess., 1957.

2. Organization for Economic Cooperation and Development, *Oil; The Present Situation and Future Prospects* (Paris: OECD, 1973).

3. Kapstein, *The Insecure Alliance*, 109.

4. The Convention was signed by the nine countries using the canal, including the Ottoman Empire, of which the Egyptian territory was part, at the time.

5. Donald Cameron Watt, *Documents on the Suez Crisis, 26 July to 6 November 1956* (London: Royal Institute of International Affairs, 1957).

6. Alfred Atherton, "The United States and the Suez Crisis: The Uses and Limits of Diplomacy," in *The Suez-Sinai Crisis 1956: Retrospective and Reappraisal*, ed. Selwyn Troen and Moshe Shemesh (London: Frank Cass & Co. Ltd., 1990), 269; Robert Bowie, "Eisenhower, Dulles and the Suez Crisis," in *Suez 1956: The Crisis and Its Consequences*, ed. W. M. Roger Louis and Roger Owen (Oxford: Clarendon Press, 1989), 197; Louise Richardson, *When Allies Differ: Anglo-American Relations During the Suez and Falklands Crises*, 1st ed. (New York: St. Martin's Press, 1996), 63; Maurice Vaisse, "France and the Suez Crisis," in *Suez 1956: The Crisis and Its Consequences*, ed. W.M. Roger Louis and Roger Owen (Oxford: Clarendon Press, 1989), 137–8; Irwin M. Wall, *France, the United States, and the Algerian War* (Berkeley: University of California Press, 2001), 33.

7. Wall, *France, the United States, and the Algerian War*, 66.

8. André Fontaine, "De Suez à Londre," *Le Monde*, 27 September 1956.

9. Mordechai Bar-On, "David Ben-Gurion and the Sevre Collusion," in *Suez 1956: The Crisis and Its Consequences*, ed. W. M. Roger Louis and Roger Owen (New York: Oxford University Press, 1989), 147.

10. Richardson, *When Allies Differ*, 25.

11. Jacques Georges-Picot, *The Real Suez Crisis: The End of a Great Nineteenth Century Work* (New York: Harcourt Brace Jovanovich, 1978), 70.

12. Organisation for European Economic Cooperation, *Europe's Growing Needs of Energy: How Can They Be Met?* (Paris: OEEC, 1956).

13. Ibid.

14. U.S. Congress. Senate, *Emergency Oil Lift Program.* 1899.

15. William Barber, "The Eisenhower Energy Policy: Reluctant Intervention," in *Energy Policy in Perspective*, ed. Craufurd Goodwin (Washington, DC: Brookings Institution, 1981), 235; Kapstein, *The Insecure Alliance*, 116.

16. Atherton, "The United States and the Suez Crisis," 270; Kapstein, *The Insecure Alliance*, 97.

17. Bowie, "Eisenhower, Dulles and the Suez Crisis," 198.

18. Some authors state that initially the administration gave some vague words of support to the British, with Eisenhower agreeing that Nasser poses a threat but rejecting military action against him, see, for example, Richardson, *When Allies Differ*, 42–5.

19. Diane Kunz, "The Importance of Having Money: The Economic Diplomacy of the Suez Crisis," in *Suez 1956: The Crisis and Its Consequences*, ed. W.M. Roger Louis and Roger Owen (Oxford: Clarendon Press, 1989), 219.

20. P. H. Frankel, "Oil Supplies During the Suez Crisis—On Meeting a Political Emergency," *The Journal of Industrial Economics* 6, no. 2 (1958): 88.

21. Alfred Michaelis, "The Middle East Conflict and Oil Supplies," *Frankfurter Allgemeine Zeitung,* 7 November 1956.

22. U.S. Congress. Senate, *Emergency Oil Lift Program.* 1812.

23. Richardson, *When Allies Differ*, 290.

24. U.S. Congress. Senate, *Emergency Oil Lift Program.* 1800.

25. Ibid., 1812.

26. Frankel, "Oil Supplies During the Suez Crisis—On Meeting a Political Emergency," 86.

27. U.S. Congress. Senate, *Emergency Oil Lift Program.* 1908.

28. Ibid., 1960.

29. "Minutes of the 91st OEEC Oil Committee Meeting," (Organization for European Economic Cooperation, 1957).

30. Richardson, *When Allies Differ*, 93.

31. U.S. Congress. Senate, *Emergency Oil Lift Program.* 1873.

32. Ibid., 1885.

33. Kapstein, *The Insecure Alliance*, 118.

34. Gerard Bossuat, "The Marshal Plan and European Integration: Limits of an Ambition," in *The Marshal Plan Today: Model and Metaphor*, ed. John Agnew and J. Nicholas Entrikin (New York: Routledge, 2004), 127.

35. Organisation for European Economic Cooperation, *Europe's Growing Needs of Energy.*

36. U.S. Congress. Senate, *Emergency Oil Lift Program.* 1927.

37. "Minutes of the 77th OEEC Oil Committee Meeting," (Organization for European Economic Cooperation, 1956).

38. U.S. Congress. Senate, *Emergency Oil Lift Program.* 1829–31.

39. Kapstein, *The Insecure Alliance*, 119.

40. U.S. Congress. Senate, *Emergency Oil Lift Program.* 1927.

41. Ibid., 1980.

42. Stephan Haggard and Beth A. Simmons, "Theories of International Regimes," *International Organization* 41, no. 3 (1987): 495.

Chapter 4
The Oil Embargo—Europe "Weakened and Humiliated"[1]

Oil use in industrialized countries had increased dramatically in the 1960s. Oil was cheap, cleaner than coal, and, in the case of Europe, geographically close. Europeans' dependence on oil has grown from 32% in 1960 up to 60% in 1970. The 1973 oil embargo, however, showed them how vulnerable their oil supplies could be. An oil embargo had been in the making for several months already. Arab leaders had discussed, on several occasions, the use of the oil weapon against Israel. Their previous successes in negotiations with the majors and the tight demand-supply balance suggested an oil supply cut might be a successful endeavor. The embargo was declared in the middle of the Yom Kippur war, in October 1973, after the first successes of the Israeli armed forces. The oil leverage was considered as a last resort to influence the Israelis and their allies. It was partially successful, as most European governments immediately declared their support for the Arab cause, and tried to distance themselves from American policy. However, it hurt producers, as well.

The embargo continued until March 1974 with disastrous effects for consumers' economies. Oil prices rose 400%, causing the worst recession since the Second World War. Inflation rates stayed in the double digits throughout the 1970s, and economic growth considerably slowed down. The main reason behind this tremendous price increase was competitive bidding and scrambling for oil. For this time, the American psychological cushion, so important in 1956, was absent. For a variety of domestic reasons, the level of oil exploration and development in the United States in the 1960s was very low, while consumption kept increasing. This left the United States with the lowest spare production capacity since the discovery of oil in Pennsylvania. As a result, the United States turned from a net exporter to a net importer, and reached its production peak. In addition to the lack of spare production capacity, there was also a shortage of refinery capacity. It was critical during the winter of 1969–1970—the coldest in many years—and became even worse the next years.[2] The availability of refined products reached its lowest level in the spring of 1973, when an oil allocation system was introduced. Naturally, no government expected anymore to receive emergency oil from the United States. This led to a heightened sense of insecurity and the accompanying panic behaviour.

Another very important factor was the disintegration of the oil companies, as a result of the nationalizations of the late 1960s and early 1970s. The companies had, partially, lost control of production, which made it harder for them to participate in oil emergency schemes. They did redistribute available supplies in 1973, however, without any coordination among themselves or with governments. Unlike in the previous two decades, they were not granted antitrust exemption, which prevented any coordination. In contrast, in 1956, a very high level of coordination was achieved through various bodies on both sides of the Atlantic, including the OEEC oil committee. This time, the oil committee gathered only five times during the entire crisis, which received very little attention.

The United States had lost its spare capacity, did not organize a crisis response, and did not provide any leadership within the oil committee. The loss of oil power was a significant factor behind the lack of American leadership. A number of other developments—both of domestic and international nature—also played a role. In the first place, in 1973 the administration was greatly weakened by the Watergate scandal. This internal havoc inevitably translated into a weakened international role. Not surprisingly, it was Secretary of State Henry Kissinger, and not President Nixon, who took the leadership in negotiations with Arab oil producers and later on, with consumers at the Washington Energy Conference. In the second place, America was still embroiled in the Vietnam War, which had become an extremely costly and unpopular endeavor. It had considerable international implications, as anti-Americanism abroad grew and questions about American leadership and power arose. The war also marked the beginning of an ever-increasing rift within the trans-Atlantic alliance, which had its culmination during the Arab embargo. The international environment, in general, had considerably evolved since the last big oil crisis. Talks with the Soviet Union produced SALT I and the Anti-Ballistic Missile Treaty in 1972, while preparations were under way for the big Conference on Security and Cooperation in Europe in 1975. On the surface, East-West tensions were subsiding. The new Ostpolitik of Willy Brandt was also bearing fruit, as a number of treaties with the Soviet Union and East Germany were signed in the beginning of the 1970s. This thaw in East-West relations reduced the Soviet threat in Western Europe, which gave America a breath of relief, as it could ease its European defenses. In contrast, in the 1950s, the Soviet threat was the primary reason for the early American intervention in oil crisis management.

This substantially different domestic and international political environment and the dire oil supply situation in the United States had serious implications for its involvement in the 1973 oil shortage. Whatever the reasons, however, the lack of US willingness and/or ability to provide leadership had dramatic consequences for the outcome of the crisis.

Crisis Background

The conflict in October came as a surprise for both Israel and the Western world, as nobody expected another war after the excruciating Israeli victory of 1967. Arabs had been preparing for war, however, ever since. Embittered by the loss of territory and prestige they suffered in 1967, all Arab neighbors of Israel were coveting a return to the pre-1967 borders. A return to the prewar status quo meant a repossession of the Sinai Peninsula by Egypt, of the Golan Heights by Syria and of the West Bank by Jordan.

Interestingly, both the Americans and the Israelis overlooked the Arab war preparations. Kissinger later shared that both countries' intelligence services have asserted that "hostilities were unlikely."[3] It is puzzling how they missed the frantic diplomatic activity of Egyptian president Sadat. If nothing else, his rapprochement with Jordan's king Hussein, who turned from an outcast of the Arab world to an ally, should have sounded an alarm. The American oil companies, active in the Middle East, were also warned of repercussions if the United States did not change its policy of support for Israel.[4]

The October war could be seen as a continuation of the 1967 dispute. For this one was never resolved. Prolonged and fruitless diplomatic negotiations took place along with a "war of attrition," which showed that the Arabs would not consent to the status-quo post. The Soviets continued supplying their allies with modern military technology and soon after the 1967 defeat, their armies had caught up. Numerous terrorist attacks, including the murder of the Israeli athletes in Munich in 1970, gave many victims. In this situation, an Arab military reprisal was foreseeable. And it happened when it was least expected—on the holy day of Yom Kippur—one of the most important holidays in the Jewish calendar. Not surprisingly, the Israeli military was on a low alert and the attack produced a huge success initially. Later, as the Israelis recovered, they were able to turn the tide in their favor. Hostilities continued until October 26. Negotiations started on October 28, and the Israelis withdrew from Egyptian and Syrian territory in the course of 1974.

The Yom Kippur War was one of those pivotal moments in Arab-Israeli history, which changed the tide of events in the region. The Arab countries, humiliated by their 1967 defeat got a vindication of some sort. However, the war also firmly established the idea among Arab leaders that Israel cannot be defeated militarily. For after the initial surprise, the Israelis recovered remarkably quickly and launched a counteroffensive, reaching deep into Syrian and Egyptian territory. The war also led to a renewed effort at Arab-Israeli negotiations, which culminated in the signature of the Camp David Accords—the first peace treaty between Israel and an Arab country. The oil market, however, proved to be the most affected issue-area, both in the short and in the long run.

The Oil Embargo

One of the most severe oil disruptions in history occurred at a critical intersection of economic and political developments. The Third World was in upheaval, détente was producing some results, but not as expected, America was wounding from the Vietnam War, and the international financial system was in turmoil after Nixon withdrew from the Bretton Woods system in 1971. More importantly, the oil market was undergoing a radical transformation—the shift from a buyers' to a sellers' market. In addition, oil consumption had spiraled up in the 1960s, while supply had remained relatively stable—a development that led to a very tight demand-supply balance. The combination of increased oil consumption in the West and producer countries' emancipation proved to be disastrous.

An atmosphere of crisis existed well before the actual shortage started. Oil experts issued numerous warnings throughout the early 1970s, which were all downplayed, since it was believed that producers needed to sell their oil, as much as consumers needed to buy it.[5] In a certain way, the 1973 oil crisis was a continuation of the 1967 one, which had only abated for five years and had come back with full force. For the Suez Canal, closed in 1967, was only reopened in 1975, and Nigeria was in the throes of a civil war until 1970. Then, in the beginning of the decade, producer governments felt empowered to request control over their oil fields and production. A long and painful process of negotiations between the governments and the oil majors began. First, official posted prices were renegotiated several times, and producers took complete control over pricing decisions shortly before the embargo. Posted prices were hitherto determined by the oil majors and were mainly intended for calculations of taxes and royalties paid to the producing governments. Market prices, based on supply and demand conditions, existed in parallel with the posted ones. Second, a number of producing countries took over oil production. Libya, Algeria and Iraq, one after the other, nationalized their oil fields. The more moderate producers negotiated participation agreements, which put an end to the era of oil concessions and the consumer-controlled market. All these changes had an enormous impact on the market and oil politics, as a whole. The era of consumers' control was over. It was a time of producers' emancipation.

The consumer side of the market was changing, as well. Oil demand in Europe had increased twofold between 1960 and 1970. In 1960, oil represented 32.5% of primary energy consumption, while in 1970—59.6%. This increase was primarily due to strong economic expansion and was covered mostly with imports from the Middle East (see table 5). The low price of oil, compared to other energy sources, and the flexibility of the oil market, compared to the coal one, were also to blame. Increased oil consumption was, also partially, a consequence of the new emphasis on the environment. Oil was a somewhat cleaner energy source than coal, and a widespread switch from one to the other started in the 1960s. In 1966, after the Thanksgiving air pollution crisis, during which the Northeastern United States found itself covered by heavy smog for a few days,

New York City switched entirely from coal to oil.[6] The following year, Congress passed the Air Quality Act, which required states to adopt air-quality standards. In general, the introduction and enforcement of clean air standards was becoming the rule rather than the exception in most Western countries. In the UK, the first clean air act was adopted in 1956, in the United States in 1963, and in Germany (the so-called TA Luft) in 1964.

Table 5. The US, France, Germany and the UK data on imports and consumption

	Imports of crude from the Middle East (% of all imports),[7] 1970	Oil (% of energy mix), 1960 and 1970	
United States	17	44.1	42.7
France	88.4	34.3	62.4
Germany	83.7	21.7	53.3
United Kingdom	84.8	25.8	44.2

Source: Data from Organisation for Economic Cooperation and Development, *Oil, The Present Situation and Future Prospects* (Paris: OECD, 1973)

Unfortunately, this period of oil consumption increase coincided with a period of constant or even decreasing levels of production. Cheap oil did not allow for new investments in oil exploration and development, and very few new discoveries were made in the 1960s. As a result, the world excess production capacity in 1973 was 500,000 barrels per day, or only 1% of free world consumption.[8] The conditions were ripe for a crisis.

Important changes were happening in the politics of the Middle East, as well. In 1973, Arab producers were more united and coordinated than ever. The change of power in Egypt—from Nasser to Sadat—led to a thaw of Saudi-Egyptian relations. Sadat's nationalistic (as opposed to pan-Arab) and anti-Soviet attitudes were more in tune with the Saudi king Faisal's policies. This reconciliation between Egypt and Saudi Arabia led to a rapprochement, in general, between the conservative and moderate Arab states. Furthermore, as a result of the 6-day war, Egypt's influence in the region declined, and the center of Arab politics moved toward oil-producing states, like Iraq and Saudi Arabia.[9] A continuation of that tendency was the establishment in 1968 of the Organization of Arab Petroleum Exporting Countries (OAPEC). Initially its membership was limited to only Kuwait, Libya and Saudi Arabia, but later it was expanded to the rest of the Arab countries. It is in the framework of OAPEC that oil ministers met in Kuwait on October 16 and 17, 1973. Their first decision was to raise prices by 59%. This was the second time producers changed prices unilaterally, without consultations with the companies. The second day of the meeting, OAPEC announced an immediate 5% reduction of oil production with subsequent 5% cuts every month until their demands were met. And those were made clear: Israeli withdrawal from the territories occupied in 1967, and recognition of the rights of the Palestinian people. In the days after the Kuwait meeting, most Arab states, led by Saudi Arabia, refined their embargo measures. The initial reduction was increased to 10%, and consumers were divided into categories, depending on their standing with the Arabs. France and the UK, which refused to allow the United Stated to use their airfields for supplies to Israel and embargoed arms shipments to both Arabs and Israelis, were put on a "friendly states" list and were guaranteed 100% of supplies. West Germany was declared neutral and got a 5% reduction. Because of their support for Israel, the United States and the Netherlands were embargoed 100%. On November 6, the Arab oil ministers, meeting again in Kuwait, decided for an immediate 25% reduction in production, and 5% subsequent monthly cuts. The biggest oil embargo ever had begun.

The selective embargo was bound to fail. The oil market was a highly connected network of transportation routes and choke points, whereby crude oil delivered in one country often ended as a final product in another. Countries with no oil resources were some of the biggest exporters of refined products. There was also a high volume of trade within Europe, which was interconnected through a complex network of oil trade deals. Germany, for example, imported 24.4 million tons of some products and exported 8.3 million tons of others.[10] The UK imported 16 million tons of products, and exported 16.4 million. Interestingly, these numbers were much lower for France, which imported only 4.4 million tons of products from other European countries and exported 9.2 million. This provided an additional shield to the French economy. France was receiving most of its oil, crude or refined, directly from Arab producers, which kept their regular supplies because of the French pro-Arab policy. Algeria, for example, exported 26.9 million tons of crude and products to France in 1970—more than to any other country.[11] This proved to be crucial during the crisis and France was the least impacted European nation. On November 16, for example, *Le Monde* reported that France was receiving 95% of its oil, while West Germany was already experiencing a reduction of about 20%.[12]

The embargo to the Netherlands produced anxiety in all European capitals. Rotterdam was the biggest port in the world at the time, and a lot of oil shipments landed there before proceeding to their final destinations. In reality, this represented a bigger problem for some North-European countries, than for the Netherlands, which satisfied a big share of its energy requirements with natural gas. Dutch officials had admitted that the embargo's long-term economic consequences could be more serious than the shortage of oil.[13] However, 25% of the OAPEC oil imported in Europe, entered the continent through Rotterdam.[14] Bonn was especially worried and for good reason. Germany was receiving 20% of its oil through the port. This might be one of the important factors accounting for the multilateral stance taken by the German government. Germany and Denmark were the only countries within the European Community to support a reallocation of oil to the Netherlands.

At the end, even the Arab-friendly European countries received supply cuts, due in part to the companies' reallocation efforts, and in part to the interdependence of the oil market. Germany got an 11.5% average reduction in its oil deliveries, France—7.3%, and the United Kingdom—0.8.[15] Production was partially offset by increases in Canada, Iran, Indonesia and Eastern Europe. Nevertheless, in December, oil production was down 7%—its lowest level during the crisis.[16]

Despite the dire predictions of most experts, the embargo lasted shorter than expected. By the end of December, it was already relaxed, so much as Germany ended its Sunday drive ban in early January. One of the reasons for this outcome was the division among producers that appeared in the beginning of the embargo and grew stronger with time. Two different positions emerged within the Arab states—the moderates (Saudi Arabia and the small Gulf states), and the radicals (Libya and Iraq), with Algeria in the middle.[17] Iraq, for example, did not join the embargo, arguing instead for a full nationalization of petroleum assets. Iraqi oil, however, also stopped flowing for a short period, after attacks on Eastern Mediterranean oil terminals.[18] Another reason was producers' quick realization that the embargo was hurting them, as well, and that it was not achieving its desired goals. The embargo was officially revoked on March 18, 1974.

Most authors and the press agree that there was not a serious shortage of petroleum on the market.[19] *The Economist*, citing Lloyd's, reported that 39% more tonnage of tankers sailed from the port of Ras Tanura in Saudi Arabia in December 1973, as compared with a year earlier. The Arabs, the article later argued, may not have cut production with the initially stated levels.[20] The same opinion was shared by a *New York Times* article, which reported tankers lining up at sea to get into ports.[21]

The economic consequences of the embargo, however, were dramatic. Oil prices rose 400% between October and January. Naturally, this led to an overall increase of prices, balance-of-payment problems and recession. After three decades of shortage of labor, European countries now suffered from high unemployment and abolished all guest worker programs. The United States was also affected. Unemployment reached close to 8.5% in 1975—the highest figure since the Second World War. As Ikenberry argued later, the energy crisis "disrupted both international markets and national economic systems" and made radical macroeconomic adjustments necessary.[22]

While the price increase was partially due to market developments prior to the embargo, competitive bidding among consumers was also to blame. Stockpiling continued during the crisis and Germany and Italy had higher oil stocks on December 15 than in previous months.[23] This "mad scramble for petroleum," as Senator Frank Church called it during a Congress hearing in January 1974, led to increased demand at a time of a very tight supply.[24] Moreover, the lack of coordination among consumers created additional market distortions, which only deepened the crisis. Given the gravity of the oil shortage and the insecurity as to how long it would last, consumers' efforts to ensure supplies is understandable. A coordinated response, however, could have alleviated some of the gravest consequences. Common policies might have taken several forms, including a joint drawdown of stocks or allocations after negotiations with the oil companies. The bilateral contracts signed between European and some Middle Eastern countries in the months after the crisis also contributed to the price increase, since these guaranteed oil supplies at exorbitantly high prices. Consumers were ready to pay anything to secure future supplies.

The United States

The United States' decline in the oil issue-area started long before the Arab embargo. The reasons were mostly of domestic nature. At the bottom of the problem was the Mandatory Oil Import Quota Program, established in 1959, so called to distinguish it from its voluntary predecessor, in force between 1957 and 1958. The mandatory program aimed at protecting domestic production, and in particular smaller, independent producers, who felt increasingly threatened by oil imports. The system remained in force for fourteen years. Quotas were set regionally, with an aggregate import limit of 9% of local demand. The Kennedy administration increased the ceiling to 12.2% of production.[25]

The program was quite successful in helping small domestic producers. In 1968, US crude oil production was 29% higher than in 1959.[26] The majors, while initially opposing the quotas, managed to adapt later and to increase profits from their domestic holdings. The program, however, also had some unintended large-scale consequences for both the domestic and world oil markets. On the home front, it led to a faster than normal depletion of oil resources. While the program maintained a higher US oil price, which contributed to the high inflation figures sparked by the war in Vietnam, it led to an oil glut and lower prices on the world market. It led to a higher level of investment in domestic oil production relative to foreign investment. It also led to an increase in global oil trade as majors needed to find other markets to sell non-US production.

These developments coincided with a period of growing consumption, which went 66% up between 1960 and 1970, partially due to the massive switch from coal to oil of electric utilities.[27] As a whole, import controls maintained an artificial level of production and imports, and delayed the natural economic adjustments of a free market economy. The growing demand and use of foreign oil made it necessary to abandon them. This was done gradually, first by increasing the amount of imported oil in 1971, and then by removing all import barriers in May 1973.

Another event that indirectly influenced the oil market was the price freeze of August 15, 1971, introduced in order to control domestic inflation, which had reached 5% in 1970. This measure was part of a packet of reforms introduced with the so-called Nixon shock, which also included a wage freeze, a 10% import surcharge and end of the convertibility between the US dollar and gold. The price of oil was included in this measure. In retrospect, this is considered to have had a very damaging effect on US energy security in the long run.[28] It discouraged domestic production, and led to a growth in demand, which was increasingly being satisfied with Middle Eastern imports.

These developments led to the gradual decline of the American dominant position in oil. As a consequence, by the time the Arab oil embargo was declared, the United States was experiencing serious oil shortages. There was already a "near-panic buying" of oil in August 1973, which brought the prices higher a month before the embargo started.[29] In fact, the measures introduced to counter the oil shortage were prompted by the domestic problem and not the Arab embargo. At the time the United States received only 17% of its oil imports or about 10% of consumption from the Middle East. The entire administrative structure, established to cope with the oil shortage worked for the domestic interest. The Foreign Petroleum Supply Committee, so instrumental in 1956, was revived again, to advise the government on local oil availability and supplies. This body was established to deal with the 1951 shortage, and at the time, participated in petroleum allocation to European allies. In fact, its main purpose was to coordinate supplies of oil to Europe.[30] It had been reactivated during the 1956 and 1967 crises with the same goal. Its change of focus to domestic supply illustrates the unwillingness or inability of the United States to serve as a supplier of last resort.

Across the Atlantic, the old emergency bodies suffered the same fate. The United Kingdom also activated its emergency petroleum committee, which, in 1951 and 1956 effectively participated in oil emergency management together with the American committee and OEEC. As in the United States, however, in 1973 this body assumed domestic functions only. Self-help, it seems, was the preferred mode of action.

The loss of American spare production capacity had grave consequences for the United States and its allies' energy security. This excess capacity has been the most critical factor in all postwar oil shortages, as well as during the Second World War. In 1951, 1956 and 1967, the increase of production in Texas proved to be the single most important remedy to the oil shortage in Europe. The absence of American excess oil "represented a major change in the underlying dynamics of politics and oil" and the end of the era of an America-led oil market.[31] It had important implications for policy coordination among Europeans, and in the crises that followed, self-help policies predominated.

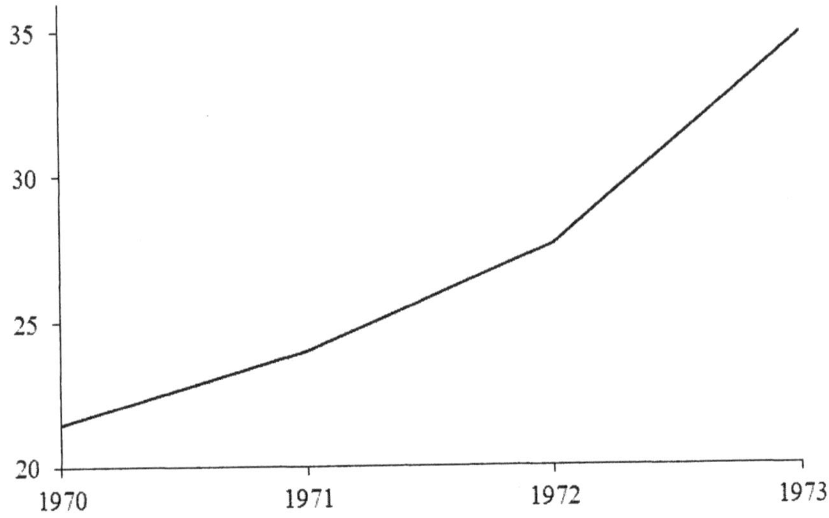

Figure 3. United States' net imports (% of consumption)
Source: Data from U.S. Energy Information Administration, "Petroleum Overview: 1949–2008," http://www.eia.doe.gov/emeu/aer/txt/ptb0501.html (date accessed 18 November 2011)

European Discord

The newly acknowledged power of oil-rich countries to use their oil as a tool of coercion produced a wedge in the Western alliance, as consumers hastened to secure oil supplies. On the European front, self-help behavior was at its highest. It took mostly two forms: bilateral contracts with producers and introduction of export restrictions. The oil embargo also caused the biggest rift in trans-Atlantic relations since World War II. According to Yergin, the Arabs' plan, right from the beginning, was to split the consumers, by differentiating among them.[32] Britain and France had learned the lesson of 1967, when only countries considered to be Israeli allies were embargoed. This time, they refused to support Israel in an effort to placate the Arab exporters. They even increased arms shipments to their Arab clients and distanced themselves from the United States, which on October 14 established an air bridge to supply military equipment to Israel. Losing their primary energy source was not an option, especially for Europe and Japan, and support for the Israeli or for anyone on their side was out of the question. With the exception of Portugal, European countries barred American planes from using their bases and air space. This refusal to cooperate with a NATO ally stemmed from the worry of Europeans to be put on the "unfriendly" list. A fear that was validated later, when Portugal was also embargoed. Later, it became clear that Turkey has tacitly allowed Soviet assistance for Arab countries to pass through its territory, while banning American military aid to Israel. All this prompted Kissinger, then secretary of state, to declare: "I don't care what happens to NATO, I'm so disgusted."[33] The European allies were also disappointed by their American ally. The United States raised the state of alert of its armed forces on October 25 without consulting them. As a *New York Times* article commented a day later, "Europeans are trying to have it both ways"—they distanced themselves from American policies in the Middle East while at the same time, complaining about US actions.[34]

In this situation of high demand, no spare capacity, and insecurity as to the Arabs' intentions, consumers took a course of *sauve qui peut* policies. Incidentally, the French, who coined that phrase, were the most zealous in pursuing bilateral deals. This might have been due to their higher dependence on oil as compared to other European countries. Germany and the UK had coal reserves, and oil took 53% and 44%, respectively, of the energy mix. For France that figure was 62%. On the other hand, France was the least affected among the three. Germany suffered because of the embargo to the Netherlands, and the UK was affected by the big miners' strike, which coincided with the embargo and paralyzed the country for several months. The reason for France's choice of bilateralism is, more probably, to be found in its tradition of Gaullist-type policies of political and economic independence. For example, unlike in the UK and Germany, the French government had a monopoly on the oil industry. During the crisis, France was the most fervent opponent of common action within the EEC or the OECD. The French government also signed more bilateral agreements with producers than any other consumer. Without a coercive leadership, the French understandably opted for bilateralism.

Although part of the Arab "friends" list, France also experienced supply cuts, due mostly to the companies' efforts to allocate oil equally. At the end of November, they notified the French government that they would reduce oil deliveries by 10 to 15%.[35] This reduction never materialized, and between December 1973 and March 1974 the availability of petroleum in France (as well as in the UK) was only about 5% lower than a year earlier.[36] Nevertheless, as soon as the embargo was announced, the French government started negotiations with producers and in January, preliminary contracts were already on the table. This process culminated in a ten-year $5 billion development agreement, including provisions for the sale of nuclear reactors to Iran, signed in June 1974.[37] Most of the agreements had extensive provisions for French investment in infrastructure, and natural resources development, as well as arms sales. Besides with Iran, in 1974, France signed agreements with Iraq, Algeria and Saudi Arabia.[38] These deals served the French interest well. They guaranteed the highly needed oil, while at the same time providing employment and improving the balance of payments. However, in terms of lowering the price of oil, the contracts had no effect, as most of them were signed at exorbitantly high prices. While they did bring some measure of security of supply for the future, they did not alleviate the most nefarious consequence of the oil shortage—high prices.

Germany, which initially insisted on a collective solution, did not wait long to dispatch emissaries to the Middle East. Not as vehemently bilateral as her European partners, Germany campaigned for an Atlantic approach to the energy problem. The bare facts might explain this difference of position. Germany was the most affected country among the three biggest economies, importing a large part of its oil through Rotterdam. It also had the lowest amount of oil stocks—54 days of consumption, as opposed to 66 for the UK and 77 for France (see table 6). Germany did not have special ties with Arab countries either, nor did it host an oil major. It needed assistance to handle the oil crisis. However, the self-help spirit of the time contrived the German government to resort to bilateral negotiations with Arab producers. For example, it signed a contract with Iran, allowing the latter to participate in its Krupp steel operation.[39] To gain more control over the oil market, the government also decided to merge two of the biggest energy companies in the country—Veba and Gelsenberg—thus, acquiring a large interest in the oil industry.

Table 6. France, Germany and the UK—stock levels, 1973

	1973 average stock level (millions of barrels)	1973 average stock level (days of consumption)
France	201	77
Germany	181	54
United Kingdom	156	66

Source: Data from Organization for Economic Cooperation and Development, *Oil; The Present Situation and Future Prospects* (Paris: OECD, 1973)

For Germany, this represented a policy shift, much more so than for France and the UK, which, generally, had more control over the oil industry. Germany had traditionally given preference to free market solutions, and if they failed—multilateral diplomacy. The energy crisis, however, showed how vulnerable the country was to outside forces it did not control, and how damaging these could be.

The UK, although less dependent on foreign oil than the others, was also badly hit by the crisis. Coincidentally, coal mine workers went on strike in November. The strike evolved to a major confrontation between the government and the minors, and continued until March 1974. The combination of oil and coal shortage greatly injured the British economy, and the government declared a state of emergency. This was the biggest energy crisis for the country since the coal shortage of 1947. Industry went on a three-day workweek, and electricity and heating suffered regular interruptions. This dire situation also hurt the British-American special relationship. The crisis marked the lowest point of this special alliance, and for a period of time the exchange of intelligence information between the two countries was interrupted.[40] The UK signed one bilateral deal—with Iran in January 1974. It provided for the exchange of Iranian oil with industrial products.

Protectionism was in vogue and another important token of this trend were the trade restrictions and price controls introduced by most European countries. For example, Belgium established export licensing, while Italy and Spain restricted exports to the United States.[41] The Netherlands and Great Britain also introduced different forms of export restrictions. These protectionist measures also hurt the United States, which imported heating oil from Europe—a highly needed fuel during the winter months. They were in contradiction with the Rome Treaty (articles 30–34), which prohibited the introduction of any trade restrictions among member-states, or other "measures having the equivalent effect." There was also an informal agreement within the OECD oil committee against these practices. Obviously, the process of European integration found itself victim of protectionism and self-help.

This bilateral trend continued after the crisis. More national companies were established and supply deals signed. Consumers had discovered their vulnerability. Kapstein argued that European behavior was "in accordance with the immediate national interest," which was expected during times of crisis. According to Prodi and Clo, the bilateral preferences of Europeans were only a short-term solution of a problem that these countries have usually solved multilaterally.[42] Future crises, however, proved that this was not a short-term tendency.

Although not covered by this study, the Japanese government's bidding for contracts is worth mentioning, since it captures so well the spirit of the time. Early in 1973, four new Japanese oil companies were created and direct negotiations with producers started. Already by the end of the year, 10% of Japanese oil was imported through government-negotiated agreements.[43]

International Cooperation: The Organization for Economic Cooperation and Development and the European Community

The oil embargo did not receive much attention within the respective organizations. The OECD oil committee hardly touched the issue, while European Community discussions were marked by deep divisions. The protectionist mood of the time brought about the biggest rift within the Community, as member-states introduced trade restrictions and refused to help the Netherlands. No coordinated emergency measures were agreed by any of the two bodies.

The OECD

The OECD was established in 1961, as the successor of the OEEC. The latter had been considered a success, having accomplished all tasks it was assigned. The Marshal Plan had been successfully executed. The European Payment Union, which had allowed for an enormous increase in trade in conditions of no currency convertibility, was abolished in 1958, after convertibility was restored. The organization had also contributed to trade liberalization in Europe through the Code for Trade Liberalization—a task now assumed by the EEC and EFTA. In short, the OEEC had successfully managed European postwar recovery, and coordination in a range of policy areas.

The OECD grew out of the idea to create a trans-Atlantic forum for economic cooperation, where states could discuss "free world problems."[44] Its membership was extended to the United States, and later to countries outside the Euro-Atlantic area, like Japan and Australia. Although, continuity was ensured, and most technical committees remained in place, the new organization was stripped of some of its previous functions. For example, trade liberalization and some other economic and monetary tasks were no longer under its portfolio. Since most of its tasks were taken over by the EEC, the OECD continued with somewhat scaled-down responsibilities, like monitoring developments, gathering statistics and consultations.

In contrast, not much changed in the work of the oil committee. It continued its regular meetings, and the only change was the membership of the United States, which was highly welcomed by the others, especially given the market influence of American oil companies.[45] The establishment of an emergency allocation scheme continued to be a high priority on the committee's agenda. The discussions, which took several years, showed how difficult interstate negotiations in such highly sensitive matters can be. They also illustrate the limits of cooperation in these areas. Lines of division appeared between bigger, more affluent countries, like France and the UK, and smaller members, for example Greece, which expressed concerns about its ability to maintain the required stocks, among other things. Countries stumbled upon issues such as which products should be included in an apportionment scheme, and in general how wide such a scheme should be.[46] Needless to say, governments were worried mainly about two things: interference in their affairs by an intergovernmental body or by other countries, and distribution or how much everybody would contribute to and receive from the common pie. According to the emergency plan adopted in 1962, 90% of available supplies were to be distributed among countries on the basis of consumption levels, the remaining 10%—allocated to the neediest, or left for particular circumstances. This scheme was never tested. During the one major oil crisis that occurred after its adoption—the 1967 six-day war—oil allocations were successfully managed by a plethora of international committees, including the OECD oil committee, but not on the basis of this plan. It remained a dead letter until the establishment of the IEA in 1974, which took over emergency planning.

In the aftermath of the 1967 crisis, the emergency agreement came under discussion again. Discussions on a new scheme took place over the course of 1972 and 1973, but a new plan was never adopted. A decision was only taken to increase the stock requirement to 90 days of consumption (from 60). Most countries were far from achieving this requirement, but still had a non-negligible amount of stocks.

It seems, at the time, there were low expectations of another oil crisis. A High Level Group, formed to study the implications of growing oil consumption in OECD countries, including "the consequences of hypothetical interruptions," concluded that there was a low risk of a supply disruption.[47] Two reasons were put forward: the number of exporting countries had increased, and the importance of the Suez Canal as an oil transportation route had significantly diminished. Moreover, exporters needed the oil income for development expenditures. Looking at different scenarios, the report concluded that only if all transit routes and three major Middle East sources were lost, the disruption would have severe consequences. Such an extreme situation was considered unlikely. The report was published in 1967—two years before Kaddafi came to power, and unleashed the wave of price negotiations and nationalizations. The possibility of producers taking over exploration and production, or acting together to increase prices was also considered minimal and was not to be "a major factor in energy planning." The Group highlighted, however that if a disruption were to occur, it would be more damaging to OECD economies than in the past and gave some recommendations to MSs on how to avoid such a disastrous scenario. Then, as now, the recommendations included higher stocks, diversification of supplies and exploration of OECD resources. However, unlike now, the report also stressed the importance of American spare production capacity for "the security of the rest of the OECD's oil supplies."[48] Since this capacity was expected to diminish, and no other factor of similar importance was seen on the horizon, the Group advised governments to consider modifications of their supply policies. Suggestions were also given for an emergency allocation system. The mention of the American spare capacity is critical. Apparently, its importance for European energy security was recognized, however not much thought was given to the consequences of its exhaustion. There was hope that the existing institutional mechanisms would help fill the void. Nobody realized, however, that for these to be effective, strong leadership was necessary, as well.

During the crisis, the oil committee met several times: three times in 1973 (October, November and December), and twice in 1974 (March and June). No emergency planning took place, and the industry advisory board—so critical in 1956—was not activated, mostly due to French opposition.[49] The Board was only instructed "to put itself in readiness."[50] At the December session, the discussion on a possible supplies apportionment scheme was deferred until the next meeting. Countries limited the crisis measures to exchanging information on demand restraint and stock levels. The United States tried again to take the lead in emergency planning and invited oil committee representatives to establish direct contacts with the American Emergency Petroleum Supply Committee.[51]

Such inter-committee meetings were essential in coordinating previous emergencies. However, they did not materialize in 1973.

It is difficult to assess from these meetings' minutes the exact reasons for their unsuccessful outcomes. According to Lieber, French and British disagreements blocked any OECD common action.[52] Before the crisis, disagreement centred on the details of the apportionment scheme—whether it should be based on consumption (the European and Japanese preference) or on imports (the American preference). After hostilities started, most consumers opposed concerted action for fear of Arab retaliation. At the October meeting several countries managed to block a statement calling for the implementation of an OECD-wide oil-sharing scheme.[53] Stobaugh argues that the oil committee played down the crisis and the evidence that a serious shortage was under way.[54]

It is also possible that the prevalent idea of the time of supply redistribution to the neediest countries was partially to blame for this outcome. Supply apportionment later lost its favor with consumers, and is currently not considered at all, although it is still an essential part of the IEA emergency procedure. Its rigidity might have prevented states from reaching at least a modicum of cooperation.

The European Community

The European Community common action or rather the lack of it, is even more interesting to study in this case. For discord appeared here, as well. The Community was the only other organization, where attempts were made to coordinate energy policies, although there was no common energy policy at the time. The EC would provide the perfect forum for cooperative action in this case, according to institutionalists—a small number of kindred states, highly interdependent, with a long shadow of the future. Moreover, the potential for issue-linkage was very high, given the high level of integration in other policy areas. This is the strategy the Netherlands attempted to use in their effort to get support from fellow EC members. It did not work, however, and even worse, some of the founding principles of the Community were violated during the crisis.

The European oil emergency system was put in place in 1968, when member-states agreed to keep stocks equal to 65 days of average consumption. As a result of the oil market turbulence of the early 1970s, the stock requirement was increased to 90 days. In addition, member-states agreed to implement several other measures, in the event of a supply disruption: consumption restrictions, priority of supply to certain groups of users and price regulation. The first component later became the first element of the IEA emergency system. In 1973, none of these measures was implemented in coordination. Member-states introduced a number of policies to tackle the crisis, including rationing in some countries, none of which was coordinated with other members. On the contrary, crisis deliberations within the EC Council were marked by disaccord.

Initially, the Nine showed a certain degree of consensus by issuing a common declaration, calling for Israeli withdrawal from occupied territories and recognition of Palestinian rights. This statement was criticized by the American administration, which accused Europeans of capitulating to Arab demands.

However, it gave Europe a short relief, as on November 19, OAPEC decided not to impose its planned 5% December oil reduction.

This initial urge to act in coordination did not last long. The decision of the Arab producers to impose a 100% embargo on the Netherlands created a rift between the latter and the other EC members, especially France. On 30 October 1973, the Dutch requested assistance from their European partners, which was promptly refused.[55] This happened several days after the European Commission president Ortoli assured the Dutch of the support of their European partners.[56] The French government was especially adamant in its position, fearing the loss of its special ties with Arab countries. This position is understandable. By reallocating oil to the Netherlands, Europeans were risking a full oil embargo, which they were trying to avoid at all cost. In addition, there was no common energy policy at the time, so member-states did not feel legally obliged to share their oil. The year, which had started as the "Year of Europe," celebrating the grand, and so far successful project of European integration, was heading toward a less glamorous end.

In the middle of November the Dutch government issued a threat to stop natural gas supplies to France, Germany and Belgium.[57] Lieber argues that this softened French position a little, since France received 40% of its natural gas supplies from the Netherlands. The Dutch government also slightly altered its position and singed onto the next EC Common Position, issued on November 20. However, no oil was shared with the Netherlands, neither within the OECD oil committee, nor within the EC. The country was embargoed until July 1974—longer than any other consumer. It was also one of the few countries, which put in place gasoline rationing—between January 12 and February 4, 1974.

This refusal to aid the Netherlands and the introduction of export restrictions, discussed earlier, represented a violation of the Rome Treaty. The free flow of goods, which was supposed to be achieved by 1970 was one of the pillars of the Rome Treaty. Europeans were in disregard of one of the foundational principles of the Community. Their bilateral negotiations with Arab producers also represented a departure from the Global Mediterranean Approach—an EC initiative, targeting most Arab countries en bloc.[58] The goal was to lay the foundations for a common policy toward parts of the Middle East. The embargo and the rift among Europeans that it created significantly delayed that process. National policies were coming to the fore again. It is astonishing how easily the integration process could be sidetracked, when national interests were threatened. Lieber has summarized European attitudes very well:

> In this sense, the energy crisis was considered a matter of national survival: oil supplies were crucial for industry, agriculture, heating, electricity, and transportation, and narrow conceptions of national self-interest received priority over European solidarity. Economic interdependence and the existence of regional integration could not—at least temporarily—override notions that oil had become a matter of national security.[59]

The Role of the International Oil Companies

The oil companies were, this time, "the only entities capable of managing the cutback and embargo."[60] They redirected oil supplies from non-Arab producers to the embargoed countries, while increasing Arab oil supplies to the non-embargoed like the UK and France. In this manner, the companies diverted about a third of their oil.[61] Unlike in previous shortages, however, companies did not coordinate action among themselves or with governments. There was not an agreed supply reallocation scheme, as in 1956, and each company acted on its own. Some sources indicate that companies did consider cooperation but decided against it, due to antitrust regulations in the United States and competition law in Europe.[62] Because of the companies' efforts, countries received similar percentage cuts in their petroleum supplies, regardless of whether they were on the Arab "friends" list or not. This is because companies did not discriminate in favor of their home markets, in spite of pressure by governments, in particular Britain and France. Without a doubt, the main reason behind this seemingly altruistic behavior is long-term profit maximizing. Companies would not risk their standing by taking political sides, breaking contracts and losing customers.

Conclusion

In 1973, Western Europeans realized how vulnerable their energy supplies could be. Oil shortages, without the security of American spare capacity and leadership could have devastating consequences for their economies. In the years after the oil embargo, economic growth slowed to an average 2.3%, down from 6.8% in 1973, and inflation was 10% for the next several years.[63] In France, the oil import bill climbed from 20 million francs in 1973 to 52.5 million in 1975.[64]

The oil embargo was the first crisis that occurred after the American decline in oil. It also proved to be the biggest postwar oil crisis, not only in terms of the physical shortage but also because of the unfavorable market conditions at the time, which greatly exacerbated the emergency. Low world spare capacity is the most important among these conditions, which also include high reliance on oil and low inventory levels. For example, in 1956 oil made up only 15% of the energy mix in Europe, while in 1970, this percentage has grown up to 60. High energy intensity at the time also meant that oil prices had a much bigger impact on overall price levels and the economy, in general. It currently takes less than half as much oil to generate $1 of GDP than it did in 1973.[65]

These market conditions, combined with the American inability to lead, produced serious consequences for the world economy. Before 1970, American leadership had consisted of excess oil capacity, early-on emergency planning, and coordination of emergency policy. All three factors were absent in 1973. During the years after Suez, and in particular in the 1960s, US oil output had declined, while consumption had increased dramatically. US wells reached their production peaks around 1970, and started producing at full capacity at that time.

Accordingly, America was now satisfying a big proportion of its oil hunger with imports. It was unable anymore to assist allies with oil in case of emergency. Coupled with domestic weaknesses of various origins, including the Watergate scandal, this translated into lack of preparation for the crisis. The Foreign Petroleum Supply Committee, so instrumental in previous crises, was revived again, but this time, only for domestic purposes. This change was significant and with grave implications.

In these conditions, European governments rushed to secure supplies and fill in inventories. They followed self-help policies, introducing export restrictions and pursuing bilateral deals with producers. Panic buying and competitive bidding brought the price of oil up 400%. No consultations took place within the newly created OECD, where the oil crisis was barely discussed. The European Community also proved helpless, and bitter disputes took place within the Council, where the Dutch—fully embargoed by Arab producers—were bluntly refused assistance. These developments stood in contrast with the Suez crisis, which witnessed the highest level of policy coordination. At the time, the European Community was still a plan on paper and the OEEC oil committee did not have emergency powers. Yet, the latter provided a very effective forum for policy coordination. In 1973, neither of the two institutions proved able to foster cooperation. As we shall see in the next chapter, the Iranian Revolution crisis in 1978 had a similar outcome. Thus, variance within the institutional variable cannot sufficiently account for variance in the level of cooperation. American leadership in oil seems to have a larger explanatory power.

After the crisis, consumers united in the new International Energy Agency. There was renewed hope that international organizations would fill the vacuum left by American leadership in oil. Soon, however, the new international setup was going to be severely tested. Whether consumers had learned the lessons of the oil embargo is the topic of the next chapter.

Notes

1. The European Commissioner for energy as cited by Robert J. Lieber, *Oil and the Middle East War— Europe in the Energy Crisis*, Harvard Studies in International Affairs, no. 35 (Cambridge: Center for International Affairs, Harvard University, 1976), 2.

2. United Press International, "Seasonal Shortage of Heating Oil Seen," *New York Times*, 23 November 1972.

3. Walter Laqueur, *Confrontation: The Middle East and World Politics* (London: Abacus, 1974), 66.

4. Robert B. Stobaugh, "The Oil Companies in the Crisis," in *The Oil Crisis*, ed. Raymond Vernon (New York: W. W. Norton&Company Inc, 1976), 183.

5. See, for example, Stephen J. Randall, *United States Foreign Oil Policy since World War I: For Profits and Security* (Montreal: McGill-Queen's University Press, 2005), 269–272.

6. Yergin, *The Prize*, 568.

7. Including Algeria, Libya and Egypt

8. In 2008, spare capacity reached again one of its lowest levels—1.8%—a situation that could easily be repeated, given the strong demand growth in Asia.

9. Bassam Tibi, *Conflict and War in the Middle East, 1967-91: Regional Dynamic and the Superpowers* (New York: St. Martin's Press, 1993), 89.

10. Organisation for Economic Cooperation and Development, *Oil.*

11. Ibid.

12. Editorial, "La France ne connaît pas de difficultés pour le transport du brut qu'elle importe," *Le Monde*, 16 November 1973.

13. Paul Kemezis, "Dutch Ask Help As Arabs Cut Oil," *New York Times*, 31 October 1973.

14. Duco Hellema, Cees Wiebes, and Gerardus Tobias Witte, *The Netherlands and the Oil Crisis: Business as Usual* (Amsterdam: Amsterdam University Press, 2004), 100.

15. Romano Prodi and Alberto Clo, "Europe," in *The Oil Crisis*, ed. Raymond Vernon (New York: W.W. Norton&Company, 1976).

16. Stobaugh, "The Oil Companies in the Crisis," 180.

17. George Lenczowski, "The Oil-producing Countries," in *The Oil Crisis*, ed. Raymond Vernon (New York: W.W. Norton&Company Inc., 1976).

18. Ibid., 64.

19. Prodi and Clo, "Europe," 101.

20. "How Scarce Is Oil?," *The Economist*, 15 December 1973.

21. Clyde H. Farnsworth, "Busy Europe and Mideast Ports Cast Doubt on Oil Cut's Effect," *New York Times*, 22 December 1973.

22. G. John Ikenberry, "The Irony of State Strength: Comparative Responses to the Oil Shocks in the 1970s," *International Organization* 40, no. 1 (1986): 109.

23. Lieber, *Oil and the Middle East War.*

24. U.S. Congress. Senate, *Multinational Corporations and U.S. Foreign Policy. Part 4: Multinational Petroleum Companies and Foreign Policy: Hearings Before the Subcommittee on Multinational Corporations, Committee on Foreign Relations*, 93 Cong., 2 sess., 1974. 1.

25. Vietor, Richard H.K., *Energy Policy in America since 1945: A Study of Business-Government Relations (Cambridge:* Cambridge University Press, 1984), 120.

26. Yergin, *The Prize*, 529.

27. U.S. Energy Information Administration, "U.S. Product Supplied of Crude Oil and Petroleum Products," http://tonto.eia.doe.gov/dnav/pet/hist/LeafHandler.ashx?n=PET&s=MTTUPUS2&f=M (accessed 13 February 2011).

28. J. P. Kalt, *Economics and Politics of Oil-price Regulation: Federal Policy in the Post-Embargo Era*, MIT Press Series on the Regulation of Economic Activity (1981); Kapstein, *The Insecure Alliance*, 157; John Kraft and Mark Rodekohr, "Crude Oil Price Controls: Their Purpose and Impact," *Denver University Law Review*, 1979.

29. Yergin, *The Prize*, 591.

30. U.S. Congress. Senate, *Emergency Oil Lift Program*. 1800.

31. Yergin, *The Prize*, 614.

32. Ibid., 608.

33. David Binder, "Kissinger Said to Express Disgust at Allies' Position," *New York Times*, 31 October 1973.

34. Alvin Shuster, "Alert Puzzles Europeans," *New York Times*, 27 October 1973.

35. Clyde H. Farnsworth, "France Is Facing Oil Delivery Cut," *New York Times*, 22 November 1973.

36. Prodi and Clo, "Europe," 101.

37. Mohsen M. Milani, *The Making of Iran's Islamic Revolution: From Monarchy to Islamic Republic*, Westview Special Studies on the Middle East (Boulder, CO: Westview Press, 1988), 109.

38. Lieber, *Oil and the Middle East War*, 32.

39. Ibid., 34.

40. Ibid., 12.

41. Stobaugh, "The Oil Companies in the Crisis," 191.

42. Prodi and Clo, "Europe," 109.

43. Kapstein, *The Insecure Alliance*, 160.

44. Richard Griffiths, "An Act of Creative Leadership: The End of the OEEC and the Birth of the OECD," in *Explorations in OEEC History*, ed. Richard Griffiths (Paris: OECD Publishing, 1997), 243.

45. "Minutes of the 116th OEEC Oil Committee Meeting," (Paris: Organization for European Economic Cooperation, 1961). The United States was an associated member of the OEEC.

46. "Minutes of the 107th, 108th, 109th and 110th OEEC Oil Committee Meetings," (Paris: Organization for European Economic Cooperation, 1960).

47. OECD Oil Committee, "The Outlook for Oil Supply and Demand," (Paris: Organization for Economic Cooperation and Development, 1967), 1.

48. Ibid., 15.

49. Clyde H. Farnsworth, "Industrial Nations Put Off Crisis Action," *New York Times*, 26 October 1973.

50. "Minutes of the 26th OECD Oil Committee Meeting," (Paris: Organization for Economic Cooperation and Development, 1973).

51. "Minutes of the 27th OECD Oil Committee Meeting," (Paris: Organization for Economic Cooperation and Development, 1973).

52. Lieber, *Oil and the Middle East War*, 15.

53. Stobaugh, "The Oil Companies in the Crisis," 185.

54. Ibid.

55. Lieber, *Oil and the Middle East War*, 14.

56. Editorial, "Le boycottage des pays arabes commence à inquiéter les Etats-Unis," *Le Monde*, 24 October 1973.

57. Kapstein, *The Insecure Alliance*; Lieber, *Oil and the Middle East War*, 15.

58. Simon J. Nuttall, *European Political Co-operation* (Oxford: Oxford University Press, 1992), 98.

59. Lieber, *Oil and the Middle East War*, 14–5.

60. Stobaugh, "The Oil Companies in the Crisis," 185.

61. Yergin, *The Prize*, 624.

62. Richard Scott, *The History of the International Energy Agency, 1974–1994: IEA, the First 20 Years* (Paris: OECD Publications and Information Centre, 1994), 107.

63. Peter Tertzakian, *A Thousand Barrels a Second: The Coming Oil Break Point and the Challenges Facing an Energy Dependent World* (New York: McGraw-Hill, 2006), 85.

64. Robert J. Lieber, "Energy, Economics and Security in Alliance Perspective," *International Security* 4, no. 4 (1980): 144.

65. Robin M. Mills, *The Myth of the Oil Crisis: Overcoming the Challenges of Depletion, Geopolitics, and Global Warming* (Westport, CT: Praeger, 2008), 30.

Chapter 5
The Energy Crisis Is "The Moral Equivalent of War"[1]

For more than two decades, disturbances in the world of oil originated in the Arab world. During Suez and Yom Kippur, the epicenter of crisis was on the Arab peninsula. More than once in this period, Iran distanced itself from Arab oil producers, increasing production to offset shortfalls. This time, however, Iran was the cause of concern.

Disturbances in Iran started as early as 1977 and continued throughout 1978, leading to the shah's hurried departure from the country in January 1979 and the return of Ayatollah Khomeini the following month. Years of painful socio-economic reforms, aimed at westernizing and secularizing the country had produced resentment among the majority of Iranians. The Revolution, which far-reaching consequences can still be felt today, changed the political dynamics of the Middle East, and in particular, Western relations with and attitudes toward the region. And in the short term, it greatly affected the oil market.

The turmoil in Iran led to a complete halt of oil production at the end of 1978. Iran, at the time, was an important exporter, second only to Saudi Arabia. The loss of its production sent ripples through the international market, and for a second time in a decade, produced a rift within the Western world. From an energy security point of view, conditions were somewhat better this time. After the First Oil Shock, European consumers had decreased oil use, and had accumulated reserves. Important conservation measures had also been introduced. However, the factors that alleviated oil shortages before 1970 were still missing—American spare production capacity, the administration's ability to quickly organize and plan the crisis response, and coordination with the oil companies. In all three respects, conditions were similar to 1973. A fourth condition—American oil consumption—was in worse shape. US oil consumption had increased 9% and domestic production had not caught up. America was now importing almost 40% of its oil requirements. In comparison, in 1956, imports comprised only 12%. The United States was no longer able to assist its allies.

The newly established International Energy Agency gave new promises for cooperation by putting in place a detailed emergency management system. However, replacing the American security cushion with an intergovernmental structure, whereby governments had to rely on their fellow members for the management of emergency oil, proved to be difficult. Let us recall that active IO crisis engagement in the past (Suez) had been possible, because of American involvement, both politically and through the supply of excess oil. These elements were missing in 1978. In addition, the position of the oil companies was further compromised after the oil embargo, as producers increased their control over the market, by obtaining majority interest in ventures and moving downstream into refining and marketing. Several European consumers and Japan created national oil companies, which further stripped the majors of their share.[2]

The US domestic environment was also less propitious in 1978 than it was in the 1950s or 1960s. Although Carter had made energy a priority for his administration, he was unable to provide the necessary leadership when crisis struck. The Energy Department, in consultations with the White House, took several months to develop a crisis response, which was rejected by Congress. A new energy policy was announced in July 1979—seven months after the halt of Iranian production. It contained broad long-term goals but no immediate solutions. In addition, the allies were barely mentioned throughout the emergency planning. This was markedly different from 1956, when their supply with oil was a priority. On the international arena—negotiations for SALT II continued, promising a new era of relaxed US-USSR relations. And while the Soviet invasion in Afghanistan put an end to détente, fears of a Soviet threat to Western Europe further abated. The United States was not only unable to lead in energy but it had also lost the rationale to do so. This unfortunate combination was going to have grave consequences for Western energy security.

Crisis Background

In 1978 Iran was an autocratic monarchy, headed by Mohammad Reza Shah. A long-standing leader propped by the United States and most other Western governments, he enjoyed considerable support abroad, but not at home. His government had embarked on a sweeping modernization program, known as the White Revolution. It encompassed all areas of economic and social activity, and ultimately led to increased literacy, improved health care, and increased standard of living for larger segments of the population. At the same time, as any other accelerated modernization effort, it deepened already existing social cleavages and created new ones. Consequently, the Revolution movement had been both praised and criticized. While appealing to the rural poor, it alienated the religious authorities and the small shopkeepers who felt threatened by the fast pace of reforms. The widening gap between rich and poor, the increasing poverty, despite the high oil income, the ever-growing inflation and the worsening living conditions in the cities, led to the formation of a strong opposition movement. The shah's authoritarian style of government contributed to the growing discontent.

In fact, the shah's authority has been contested by different groups of Iranian society, ever since his Western-supported assent to power in 1953. Popular protests became a usual sight on the streets of Tehran. Ayatollah Khomeini, who, on several occasions, found himself in the center of events, was arrested and sent in exile in June 1963. He remained one of the leaders of the opposition until his return in January 1979, uniting different groups of society with his religious, anti-Israel and anti-American rhetoric.

Ironically, the opposition movement got invigorated by the shah's political liberalization program introduced in 1977. Faced with growing protests, and encouraged by the new American agenda of support for human rights, the government put forward important reforms, including freedom of political associations and demonstrations, a free elections bill, and a code to deal with royal family corruption.[3] These reforms encouraged more open opposition and, in March 1977, a wave of open letters and petitions against the government ensued. This movement slowly grew into demonstrations, riots and strikes, culminating in the so-called Black Friday of clashes between police and protestors in September 1978, and a peaceful demonstration of over a million people in Tehran in December.

The shah, already terminally ill with cancer left Iran on 16 January 1979, admittedly to seek treatment abroad. The Regency Council that he left made a last effort to establish order and remain in power, by introducing another round of reforms, among which a release of all political prisoners, a halt of oil sales to Israel and a review of all important foreign contracts.[4] The days of the Pahlavi dynasty were counted, however. Khomeini returned from exile on February 1 and shortly after established a new government.

The Iranian Revolution had long-term consequences for domestic, regional and international politics. It led to the establishment of the first modern theocracy in the world, still in place today. The regime, based on Islamic law, became a firm supporter of Islamic movements across the world, as was the case with Hezbollah in Lebanon. On a regional level, it dramatically changed the power balance and the political dynamics of the Middle East. Iraq, hoping to benefit from the postrevolutionary chaos, invaded Iran in September 1980, unleashing a war that would last for eight years. The American twin pillar policy, relying on Iran and Saudi Arabia to "police" the region and represent US interests there, came to an abrupt end. Deprived of one of its closest allies, the United States started a vast campaign of military buildup in the region, significantly improving its power projection capabilities in the course of the 1980s. These would prove critical during the First Gulf War of 1990–91. It was in the oil arena, however, that the Revolution, and the subsequent oil shortage had its biggest immediate impact.

The Oil Shortage

In December 1978, oil became scarce again. Consumer countries had not yet completely recovered from the first oil shock, and yet another one was about to happen. As part of the protest movement against the shah, in September, workers in a large Tehran refinery and a petrochemical complex went on strike. They were followed by a strike of National Iranian Oil Company (NIOC) at the main center in Ahwaz. Slowly, the strikes spread through the entire oil industry, leading to a reduction of 27.2% of production at the end of October.[5] After three months of strikes and disturbances, all Iranian oil exports halted on December 26, throwing world markets into panic, and seriously threatening consumer countries' economies. Iran, at the time was the second largest exporter of oil in the world after Saudi Arabia. It produced 5.5 mbd, 4.5 of which went abroad. Iranian exports accounted for 10% of noncommunist consumption. Germany counted on Iranian oil more than any other Western country—it provided 12.5% of its oil needs. For France, the figure was 9.5%. The United States imported only about 5% of its oil from Iran, which was its sixth supplier, and was, thus, less affected than other consumers.[6] Great Britain had turned from an oil importer to a major exporter in the course of the 1970s.

The Iranian crisis coincided with several other developments that affected the oil market. In July 1979, the Nigerian government nationalized all BP possessions in the country. Shortly after, the company suspended crude deliveries to most of its third party customers.[7] Three months later, on 4 November 1979, a group of students attacked the United States embassy in Tehran, taking the entire staff hostage. The affair evolved into one of the longest hostage crisis ever, with 52 members of the embassy staff remaining there until January 1981.

November, in general, proved to be an eventful month, when two weeks after the embassy attack, a terrorist organization scaled a siege of the holy site of Mecca in Saudi Arabia. These events worsened consumers' fears of a deepening oil crisis, and greatly increased the uncertainty on the market, leading to what has been dubbed the "minipanic" in November.[8] And as if political turmoil was not enough, nature intervened, as well. In August, storms in the Persian Gulf reduced Iranian exports to less than 1 million barrels a day—down from more than 3 million.[9] All these events created an almost permanent atmosphere of supplies insecurity.

For a moment in the beginning, however, it seemed that the Iranian shortfall would be short-lived, and with increased production elsewhere, a full-blown crisis would be averted. Saudi Arabia was able to boost its production to 10.5 mbd—up from its self-imposed ceiling of 8.5 million b/d—in less than a month. This remarkable feat was not so difficult to accomplish at the time, since the Saudis possessed 3 million b/d (3% of world consumption) of surplus production capacity. The trouble was that the rest of the world surplus production capacity was concentrated in Iran. However, with smaller increases elsewhere most of the shortfall was covered. Kuwait, for example, expanded production with 500,000 b/d, followed by smaller increases in Venezuela and Nigeria. Taking all that into account, the actual shortage in the first quarter of 1979 was 2 mbd.[10] This amounted to only 3% of the 1978 average world consumption of 64 million b/d, and should not have represented a major market disruption.[11]

There were two more reasons to keep business as usual and not to panic. In the first place, oil stocks were at one of their highest levels ever, and considerably higher than in 1973. France, Germany and the UK had on the average stocks equal to 86 days of consumption. The United States lagged slightly behind, but was still better off than in 1973 with an increase of stocks from 58 to 68 days of consumption. In the second place, with the notable exception of the United States, which was using 9% more oil than in 1973, consumption was lower in all consuming countries. France, for example, had reduced the use of oil by 8%, while in the UK, the reduction was of 20.7% (see table 7). Conservation and the recession provoked by the First Oil Shock, were the main reasons for this decline. Overall, the world in 1978 was better prepared to face a shortage than in 1973.

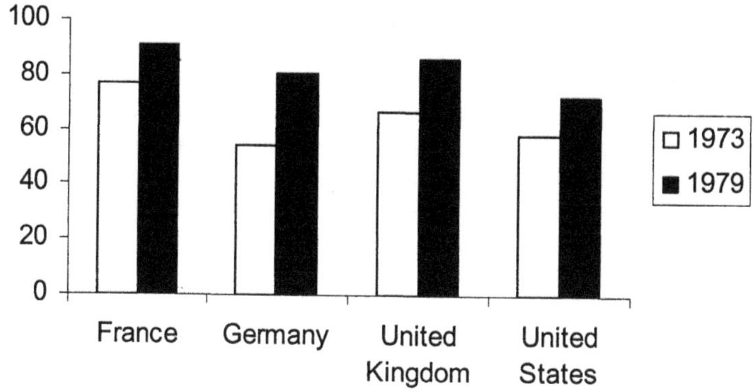

Figure 4. Stocks (days of consumption)
Source: Data from U.S. Energy Information Administration, "International Petroleum Stocks," http://www.eia.doe.gov/emeu/international/oilstocks.html (date accessed 13 March 2012)

Table. 7 Average consumption (thousand barrels a day)

	France	Germany	United Kingdom	United States
1973	2601	3324.37	2341	17307.68
1975	2252	2957	1911	16322
1977	2294	3212	1905	18431.42
1979	2463	3372.68	1971	18512.54

Source: Data from U.S. Energy Information Administration, "World Petroleum Consumption, 1960–2007," http://www.eia.doe.gov/emeu/aer/txt/ptb1110.html (accessed 17 December 2011)

Iranian production started recovering in March but stayed well below its pre-crisis level, with 2.4 mbd, of which 1.6 million for export. Increased production in other countries, combined with conservation and lower consumption, compensated for this loss. The market settled completely down in the beginning of 1980. By the summer of the same year, there was an excess supply of 3 mbd on the international market.[12] The doubling of oil prices produced by the Second Oil Shock and the recession that ensued led to a considerably lower oil demand in the first half of the 1980s and an oil glut in 1985.

Despite the seemingly more favorable conditions than in 1973, consumers again rushed to buy oil and build up reserves, which, unintentionally, exacerbated the crisis. The scramble for supplies continued well into the summer and fall of 1979. It was fueled by the widely spread perception that demand would keep increasing in the future. In fact, as a result of conservation and poor economic performance, demand dropped considerably in the next decade. Severe oil disruptions have happened twice in the course of five years, and they would probably happen again in the future, governments and businesses thought. This conviction prompted actors to build inventories at a time of short supplies.

As in 1973, first and foremost, this reflected on the price of oil. As early as November 1978, the European spot market registered prices 10 to 20% above official prices.[13] This increase did not impact the market immediately, as most oil was still delivered on the basis of long-term contracts. However, it did lead to a higher price of oil overall, as contracts' prices were set on the basis of spot prices. As a consequence, in the last week of December, OPEC raised its official price by 14.5%.[14] At the same meeting, a decision was taken to increase prices in four installments throughout the next year. An even more critical move was OPEC's decision in March 1979 to allow members to impose surcharges on oil as they wish, benefiting from the shortage and rising prices. All producers, with the exception of Saudi Arabia, which was again advocating moderation, introduced surcharges. In addition, Iraq declared that its surcharges would match the highest premiums charged by other producers.[15] This move started a leapfrogging spree, and contributed greatly to the price rise. In addition, the amount of oil sold on the spot market increased considerably, as previous engagements for oil delivery were broken, due to the shortage. The spot market, which until then had represented only 8% of oil trade, became the center of activity. And due to the recent disintegration of the oil industry, a relatively large number of actors were involved in the bidding—the oil majors, refiners, independent oil companies, state oil companies, and governments, among others.

One of the most important reasons behind this panic-induced behavior was the disruption of contracts between oil companies and producer/consumer governments. The First Oil Shock and the preceding nationalizations of oil production have led to a deep oil market restructuring. The system of concessions, which had been in place for almost a century, was replaced by a two-tier system of long-term contracts between the multinationals and producer governments on one side, and between the former and consumer countries, on the other. In addition, companies, depending on their oil positions, traded oil among each other. While not all companies were affected by the halt of Iranian production, the various actors on the market were so interconnected that one's losses reflected on everybody else, and a sort of domino effect ensued. The biggest losers were British Petroleum, with a 40% loss of its supplies and Shell—14%.[16] As a result, BP canceled its contract to supply Exxon with 350,000 b/d of crude, which, in turn, had to cancel contracts amounting to about 400,000 b/d.[17]

But most importantly of all, uncertainty was a major factor again. Nobody could predict how long the shortage would last and how far the Iranian upheaval would spread. For had it spread to other Middle Eastern countries, the oil crisis could have turned into a real disaster not only for consumers and the oil market, but for the whole world. There was no Western country that could replace the security cushion provided by the United States prior to 1970. In addition, fears were fueled by the 1973 experience. The severity of the shortage and the recession it brought still haunted Western policy makers.

Overall, from the beginning of the crisis until 1981 prices increased with 150%, mostly due to panic buying and hoarding of oil. The latter also led to an artificially increased demand of 3 mbd above normal demand. The actual shortage was then 5 million b/d. The abrupt rise in oil stocks worldwide is very indicative of the spirit of the moment. World oil reserves increased from 4.3 billion barrels in the first quarter of 1979 to 5.1 billion barrels a year later. Once again cooperation could have averted these developments, and once again, it was overshadowed by panic bidding.

The United States

The year 1978 saw the United States even more dependent on foreign oil than it was five years ago. To be fair, some lessons from the First Oil Shock had been learned. However, as soon as prices settled down, demand renewed its upward trend, and conservation measures came out of vogue. Despite all the efforts to increase domestic production, between 1973 and 1978 oil imports increased 32%.[18] And unlike in most other consumer countries, consumption grew as well—with as much as 7%. In comparison, France had reduced consumption by 7.4%, Britain—by 17.2%, and in Germany it had remained almost the same. In two important respects, the United States' and Europeans' oil security had improved—the establishment (in the United States) or increase (in France and Germany) of petroleum reserves, and diversification of imports. The United States had strengthened its relationship with Saudi Arabia—one of the pillars of the American Gulf security strategy. France had improved bilateral relations with Iraq and by 1978 was importing considerable amounts of Iraqi oil in exchange for armaments and nuclear technology. According to Lieber, at the time, this was the most successful French oil policy.[19] In addition, France and Germany had embarked on wide-scale nuclear energy programs, while also encouraging the use of coal and natural gas through subsidies and tax deductions.

The inter-crisis period also saw the introduction of wide-scale energy conservation measures. The United States adopted a national maximum speed limit of 55 miles per hour in 1974 and the Corporate Average Fuel Efficiency (CAFÉ) standards in 1976. Similar standards were adopted in Germany, as well. In general, all countries diversified oil imports as far as possible, and attempted to reduce the share of oil in the energy mix.[20]

Despite these changes, crisis behavior remained substantially unchanged from 1973. Most governments hurried to buy spot oil and build up reserves as soon as the crisis started. Attempts at coordination came late, after prices had already dramatically risen. All consumers increased their stocks during the crisis—a counterproductive but understandable behavior in times of shortage. In the United States, for example, crude stocks in May 1979 were 14.7% higher than a year before.[21] Stocks replenishing continued even after the Tokyo decision of June 1979 not to do that any longer.

There was also a renewed drive to negotiate bilaterally with producers. For example, France signed a new deal with Iraq for the supply of 100,000 b/d of crude in addition to what had already been contracted after Iraq Petroleum Company's nationalization in 1972.[22] The German foreign minister visited most important oil producers in the summer of 1979 and negotiated several bilateral deals, notably with Iraq and Libya.[23]

Albeit these similarities, a number of policy changes have taken place in the course of the 1970s, which reflected on consumers' behavior during the crisis. For example, there was a profound difference between French attitudes in 1973 and in 1978. In the latter case, the French were much more supportive of cooperative measures. This was partially due to the change of government in 1974 from Pompidou to Giscard d'Estaing, who broke with the Gaullist tradition and championed more cooperative policies both with his European partners and with the United States. Lieber argues that this transformation was also due to the realization that French energy policy did not stand alone and was influenced by decisions elsewhere.[24] There was also a process of learning, as French bilateralism evidently did not help much during the first oil crisis. This new attitude was succinctly outlined in a statement of President Giscard d'Estaing in June 1979: "There exists no solution to energy problems apart from solidarity and cooperation."[25] The new role of France was illustrated at the first meeting between EEC and OPEC, where the Community was represented by the commissioner for energy and the French energy minister.

The second big departure from 1973 had taken place in the United Kingdom, which had become a producer rather than a consumer country. North Sea oil started flowing in 1975 and four years later, Britain was reported to have achieved self-sufficiency.[26] Nevertheless, the country kept importing oil in order to avoid a refinery conversion, considered too expensive.[27] Deese and Nye contend that Britain kept its "consumer mentality," which allowed it to continue emergency planning with consumers within the IEA and the EEC. The record shows a somewhat different dynamic, however. Bitter discussions took place within the EEC, as other member-states sought a special treatment and increased supplies of North Sea oil, which Britain refused. Similar disagreements were seen within the IEA.

Unfortunately, there is no access to archival records of the International Energy Agency deliberations or government documents. Unlike in the 1950s and

1960s, there is no clear evidence that US leadership was essential in forging coordinated policies. There is one certainty, however—that neither the United States nor any other country assumed the role of the leader this time. In Kapstein's words "the United States proved powerless to overcome the situation."[28]

The United States was not able or willing to lead for several reasons. In the first place, it lacked the oil resources needed to serve as a supplier of last resort. Although the Trans-Alaskan pipeline had started transporting oil to the lower 48 in 1977, imports had kept increasing, and spare production capacity was close to zero. In the second place, and for reasons outside the scope of this research, the administration reacted slowly to events. Ironically, the Carter administration, which put in place the Department of Energy in 1977, and made energy a domestic priority, found it difficult to master the resources and consensus necessary for an effective response to a crisis of this magnitude. The presidential office started discussions on a crisis response plan in early February. The plan focused on voluntary conservation measures and fuel-switching in industry. It took several months for the plan to materialize and was officially revealed by Carter in his famous speech in July 1979. In the meantime, in April, the House voted down a Department of Energy request for authority to prepare a standby rationing plan, including a number of mandatory conservation measures. The entire crisis period was marked by deep divisions and frictions between the White House, Congress and different departments, culminating in the resignation of a number of cabinet members in July 1979, including the energy secretary James Schlesinger. These internal weaknesses translated into a lack of leadership internationally.

The contrast with the crises in 1951, 1956 and 1967 is apparent. In the first two cases, the administration had an elaborated plan of action before the actual start of the crisis. In 1956, for example, a reallocation plan was available at the end of August—three months before oil stopped flowing. In 1967, an emergency program was adopted a week after the start of the embargo, and the United States took a leading role in the re-distribution of supplies coordinated by the oil committee.

There was no plan of action in 1979. The DoE plan prepared in February hardly included any references to cooperation with European allies. According to press reports, there was not much discussion about assistance or coordination with Europeans during the interagency negotiations on the plan.[29] On the contrary, a rivalry arose, as it became clear that oil companies have been diverting oil from the United States to European markets because of higher prices there.[30] This is in stark contrast with the 1950s and 1960s, when Europe's supply of oil was given a priority. Another controversy arose in June, when the United States introduced a 5 dollar per barrel subsidy for certain petroleum imports, without informing its European partners. This measure produced a wave of protests from fellow consumer countries, exacerbated the climate of bidding, and ultimately led to an increase of oil prices with additional $5.[31]

Governments were, in general, opposed to sharing their oil. Canada, for example, vehemently protested the withdrawal of oil from its market because of its domestic production.[32] This action was interpreted as government's unwillingness to share oil with other consumers. The US government also expressed concern at the companies' oil allocations, which were said to increase the shortage in the country from 2.5% to 10–15%.[33] In fact, the *New York Times* cites Carter saying that if the IEA oil-sharing agreement were activated, the US oil shortage would increase 4%. All these disagreements reflected on the negotiations within the international organizations.

International Cooperation: The International Energy Agency, the European Community and the Tokyo Summit

International organizations reacted slowly to the crisis. The IEA and the EC both took their first crisis actions in March 1979. This is certainly expected, given the intergovernmental nature of these institutions, which makes it very difficult to organize emergencies and respond to unforeseen events. One of the main tasks of the IEA, however, was to coordinate responses to oil shortages, which usually happen unexpectedly. The EC, on the other hand, had a more integrated structure, with regular Council meetings, and binding authority. As expected, it introduced stricter and more detailed measures to counter the crisis. However, these came late, and could not avert the steep price rise. Moreover, member-states implemented them only partially or not at all.

The International Energy Agency

The International Energy Agency was established in the aftermath of the First Oil Shock. It grew out of Kissinger's multilateral approach and his efforts to organize a united consumers' front. Its establishment proved to be a challenging endeavor, even after the biggest oil shortage ever, and to a degree, as a result of it. For Europeans, energy security was enhanced not with the United States, but without it. France was especially skeptical of Kissinger's idea, and insisted on a Euro-Arab dialogue and improved bilateral relations with producers. Europeans, as a whole, came divided to the Washington energy conference in February 1974. Britain and a few others supported the French position of bilateralism. Germany preferred a common front under American leadership. According to Kapstein, a final resolution was achieved due to American linkage of energy to other issues, most notably the threat to withdraw troops from NATO.[34] The author points that this linkage of energy and security showed the weakness of American leadership. In comparison, in the 1950s and 1960s, the United States was able to achieve accord within the alliance without the use of negative sanctions.[35]

The Washington conference established an Energy Coordination Group, which met between March and September 1974. Its work culminated in the creation of the International Energy Agency in November 1974. The OECD oil committee, which had been responsible for emergency management for the last two decades, was suspended shortly after. With the exception of France, Iceland and Finland, all members of the OECD became IEA members. France kept its line of bilateral oil policy, and refused to participate in a US-dominated organization, as the IEA was considered at the time. It did participate indirectly in IEA deliberations, through the EEC, which was a full-fledged member. Iceland and Finland imported most of their oil from the Soviet Union, and feared eventual retaliation. France became a member of the IEA in 1992.

The Agency was the first international organization dedicated solely to energy. Its tasks were divided among four directorates, focusing on emergency management, long-term cooperation, the oil market, and relations with producers and other consumer countries. Not surprisingly, emergency management, based on the International Energy Program (IEP), was considered to be its most important function in the beginning. The Program consisted of a three-tier emergency plan, which, with slight amendments, is still in force. While coordination was considered important, national responses made up the first component of the system. In the first place, states had to hold oil reserves equal to 90 days of consumption, which they were expected to draw down in case of emergency. Second, the agreement provided for a coordinated reduction of consumption in times of supply shortage. Each member was required to have a program of demand restraint measures at all times. And third, after reserve drawdown had reached 50% of members' holdings, allocations would set in. The allocation principle was to be decided on a case-by-case basis, and thus involved negotiations among member-states and among them and companies. But for the whole process to kick off, there had to be a 7% reduction of supplies within a member-state or a group of member-states.[36]

The advantages and shortcomings of this system of emergency management have been discussed at length elsewhere. Several points need attention here. First, the 7% threshold had been quite ambiguous. It is sometimes difficult to establish the exact amount of oil taken off the market in a particular country given the complexity of the industry and the number of actors involved. There is no guarantee that if this level of shortage were reached in one country, other members would agree to activate the scheme. This is exactly what happened in 1979. In addition, before the 7% mark is reached, governments can unilaterally act to counter the shortage, which could, in effect, precipitate the 7% shortfall. Furthermore, the negotiations, envisaged by the scheme, are usually time-consuming, and time is a highly needed resource during a crisis. The absence of France from the organization—one of the biggest oil consumers and an important actor in European and trans-Atlantic politics, also diminished the value and effectiveness of the scheme. In short, the obligation emanating from this Program could not be enforced in conditions of self-help and anarchy, especially when states' survival is at risk. After all, the International Energy Agency is an intergovernmental institution, with no enforcement mechanisms.

These shortcomings have been taken into account, and currently the emergency management system, implemented by the IEA looks somewhat different. Provisions have been included for a pre-crisis situation of less than 7%. Procedural reforms designed to allow the agency to respond in a timelier and more flexible manner were also adopted. With the years, the IEA has reduced its reliance on allocation and strengthened the other components of the emergency response system. The joint release of stocks is now considered the dominant course of action. More importance is also accorded to market forces, as oil markets in most countries have been deregulated, and the spot market has increased in importance. Oil allocations now are seen as a last resort and there is a general understanding that they are unlikely to work as envisaged.

The Second Oil Shock happened four years after the establishment of the 7% system, and was its first test. This was the most elaborate emergency scheme so far. Did it succeed in alleviating the shortage or its consequences? On a first reading, the answer would be negative given the 150% price rise caused by the uncoordinated panic buying, and the recession thereafter. Indeed, the Agency did not contribute much to a coordinated policy. There was very little cooperation during the crisis and it took place outside the IEA. The truth is more complex, however, as the Agency did take some measures, albeit too late and too few.

In mid-November, after the first dramatic price increase, IEA director Ulf Lantzke assured member-states that lost Iranian oil would be made up by increased production elsewhere. Naturally, this reassurance was not sufficient to allay consumers' fears. It took, however, several more months for the Agency to consider more serious action. The IEA Governing Board met for a first time in March 1979, or roughly two months after the complete halt of Iranian production. Speculative bidding and stockpiling were happening as early as November, and prices had already considerably increased. A similar delay occurred in 1991, when an emergency plan was put in operation five and a half months after the oil shortage was reported. According to the IEP, an emergency meeting of the Governing Board was to be convened 2 to 6 days after an emergency is found by the IEA Secretariat. In 1979, however, the Secretariat "did not immediately find the need to take responsive measures."[37] The Agency's director Ulf Lantzke declared in late January that there was "no problem for February and no major problem for March" but a meeting was scheduled to assess the Iranian situation.[38] This is an astonishing statement given that Iranian exports have already stopped flowing at the end of December. The Secretariat did request information from companies and governments, as a first step toward evaluating the situation. As a result of this, at the March meeting, members pledged to reduce demand by 5% of the 1979 projected levels, to stabilize the market.[39] The first-tier part of the emergency plan was put into action. The reduction method and implementation was to be decided individually by members, based on their patterns of consumption and supply. According to Scott, the absence of precise targets, commitments and monitoring mechanisms reduced the effectiveness of this decision.[40] Shortly after the March meeting, Sweden and Denmark declared an 18% shortage and requested the activation of the allocation scheme. The Secretariat took three months to study the issue, and decided that the Swedish problem arose because of some peculiar conditions in the country. It identified several oil companies, willing and able to help ease the situation, but by that time the Swedish government had taken measures on its own. For the rest of the crisis period, the IEA limited its activity to monitoring developments and gathering information, making minor improvements in the data system.[41]

The cases of Sweden and Denmark deserve special attention. Clearly, they were refused the assistance they should have gotten under the IEP rules. According to the IEA, the shortage across members never exceeded 4%. This might be true on the average, but some countries, notably Sweden and Japan did experience a higher than the threshold shortage. The IEP rules stipulate that "Whenever any Participating Country sustains or can reasonably be expected to sustain a reduction in the daily rate of its oil supplies … by an amount exceeding 7% of the average daily rate of its final consumption during the base period, allocation of available oil to that Participating Country shall take place in accordance with articles 8 to 11."[42] Not activating the allocation system was clearly in contradiction with the IEA rules. In comparison, in 1956 and 1967, allocations did take place, even though in the first case, an emergency program did not exist. They were organized and led by the US authorities, with the assistance of the OEEC/OECD oil committee. In 1978, US leadership was lacking for a variety of reasons. Whether it was the weakness of the Carter administration, as some have argued, or the lack of surplus capacity, the United States did not support a concerted effort this time. As an editorial in the *Petroleum Economist* commented—it is not very likely that the 7% threshold plan would be activated, because the United States has indicated it was "unwilling or unable to do its part."[43] The article went on arguing: "while the IEA serves as a valuable gatherer and dispenser of information about the world's oil industry, it has become increasingly apparent that in a real oil crisis it is powerless to override the shorter-term objectives of governments."[44]

The IEA emergency system was doomed from the start. While it was devised with the purpose of alleviating an oil shortage, it did not take into account two important factors, which usually have a more devastating effect than the shortage itself. In the first place, the system did not address immediate crisis behavior, which is usually uncertainty and panic-driven. This behavior exacerbates the crisis and is usually responsible for its long-term consequences. Second and more important, potential economic damage was not included in the criteria for the scheme's activation. Evidence from previous crises shows that the physical shortage of oil is less damaging than its economic consequences. The price rise, provoked by competitive bidding leads to high inflation and unemployment, and ultimately—recession. None of the measures taken by the Agency alleviated the market pressure in the short-term and stopped the price rise. This is due, to a degree, to its slow reaction—almost four months after the halt of Iranian oil flows. It is difficult to know whether this delay was provoked by the slow workings of the Secretariat, which is supposed to acknowledge the shortage before an emergency meeting is convened, or by the member-states, which might have been unwilling to work together in a situation threatening their oil supplies. In any case, cooperation within the IEA did not ensue.

The European Community

The EC adopted its first oil emergency plan in the aftermath of the 1967 crisis. Its measures paralleled the IEA's provisions, in some respects, and complemented them in others. The main element of the emergency scheme was an obligation to hold oil reserves equal to 65 days of consumption. Later, as a result of the First Oil Shock, several additional measures were adopted. They were very much responsive in character, influenced by the 1973 events and did not introduce any major novelty.

The Community's measures were based on a Commission's investigation into the role of oil companies in 1973, which highlighted the damaging effects of uncoordinated action.[45] In particular, the report focused on two types of policies—national oil price controls and oil export licensing—both introduced at the height of the Yom Kippur crisis. The Commission found that these measures distorted trade, and exacerbated the consequences of the oil shortage.[46]

As a result, in November 1977, the Council adopted an emergency scheme, which introduced an important supranational element in emergency management. According to the directive, in case of a supply shortage, the Commission could take a decision for consumption reduction in member-states by up to 10%. The petroleum products saved as a result of this reduction were to be shared among member-states. The directive does not provide any details as to the extent of the shortage that would trigger these measures or the method of sharing.

This decision attempted to secure a concerted Community response to a future supply crisis. In one respect, however, the EC actually relaxed its rules, so as to enable member-states to shield themselves individually from a crisis. It allowed for the introduction of licenses for oil traded among MSs, which otherwise are illegal under the Treaty of Rome. This decision was prompted by the fact that most EC members introduced trade restrictions in 1973, even though they were not supposed to do so under EC rules. There was also a proposal to oblige member-states to continue exporting oil produced in the Community at some average level. Understandably, this idea was rejected by the UK.[47]

A number of Council decisions, aimed at easing the shortage, were taken during the crisis. In March, member-states decided to limit their oil imports to the level reached in 1978, that is, 5% on the average.[48] This commitment was similar to the one taken by the IEA Board, however EC members went a bit further and agreed concrete import ceilings on a country-by-country basis. In May, a discussion took place on eventual control of the spot market in Rotterdam, which was considered the main culprit for the price rise. This measure was opposed by Germany and was never adopted, but a decision was taken to monitor spot prices. A Commission communication from that period mentioned that the main line of disagreement among member-states was whether to rely on market forces or to impose more government controls. Germany and France were the leaders of the two opposing factions.

The apple of discord within the Community this time was Britain's North Sea oil. Several times throughout the crisis, France and Germany suggested an increase of British oil imports to fellow member-states to offset the shortage, in a long-term supply guarantee.[49] A proposal was even made to consider British oil a "common resource" of all members, and to put it under supranational authority. Naturally, the British government opposed such a move.[50]

Given the high interdependence of European economies and the frequency and nature of member-states' interactions within the EEC institutions, one would expect a higher level of emergency cooperation within this organization. In reality, some coordination did take place, but it hardly exceeded the measures already adopted by the IEA. Member-states undertook unilateral actions as early as November, without consulting their EC partners. The first Council meeting took place in March 1979. In addition, most countries did not comply with the EC decisions and continued pursuing unilateral policies. In Germany, for example, oil imports rose 5.5% in 1979, despite both IEA and EEC decisions for reductions of imports.[51]

The lack of consultations among EC members can be explained in only one way—unwillingness to coordinate in times of crisis. The European Community has a permanent Council secretariat and a Commission with supranational functions. Meetings on different policy areas take place very often, and energy ministers meet at least twice per year. So do heads of states. These settings allow for quick consultations in cases of emergency. When the oil supply was threatened, however, MSs opted for the self-help approach.

The Tokyo Summit

Interestingly, the most important crisis decisions were taken outside the framework of the EEC or IEA. In June 1979, a G-7 summit took place in Tokyo. It was conceived as an economic meeting, aimed at discussing the most pertinent economic and financial issues. It turned into an energy summit, leading to several critical decisions. The big seven pledged to impose country-specific ceilings on oil imports and maintain them through 1985. This was, in a way, the EEC decision, taken earlier, extended to other countries. In addition, the G-7 members agreed not to supply national oil reserves anymore and to consult each other before resuming open market purchases. These two decisions were the most reasonable actions taken so far by the international community. Had they been taken seven months earlier, the crisis could have been much less severe. Not surprisingly, Carter declared the Tokyo results "superb," while European Commission president Roy Jenkins thought of them as "the first concerted response to the new energy crisis."[52] It is difficult to know whether these decisions were observed. There are press reports indicating the contrary. This is probably correct, as there are indications that most consumers' reserves kept increasing. However, the fact that the most important decisions were taken outside of the respective international bodies is indicative.

Conclusion

The 1978 oil crisis had an even more devastating effect on consumers' economies than the crisis five years earlier. It exacerbated the recession trends of the post-1973 period. Inflation increased again and reached 10.8% in France and 12.3% in the United Kingdom in 1979. Inflation-averse Germany had a rate of 4.5%. In 1982, worldwide economic growth reached one of its lowest points since World War II—0.8%. This dire economic situation was partially caused by high oil prices, due to competitive bidding and hoarding of supplies. In 1978, for a second time in five years, consumers could not form a coordinated crisis response. The 1973 experience had, obviously, not been a learning one.

This case clearly lends support to the hegemonic stability thesis. Despite a variety of conservation measures, adopted after the First Oil Shock, in 1978, the United States consumed more oil and was more dependent on imports than five years earlier. Consumption had risen 7% in the inter-crises period. Needless to say, domestic production was at full capacity. In this situation, the administration focused on countering the domestic supply shortage, and allies were not given any consideration in the proposed emergency plan. In any case, the measures, approved in July 1979, focused on the crisis' long-term effects, and not on its immediate consequences. The Carter administration had difficulties handling even the domestic shortage.

Across the Atlantic, panic-driven behavior was the norm again, and consumers eagerly built up inventories and signed bilateral deals. The panic started in November 1978 and, with small intermissions, continued until the end of 1979. The two international bodies reacted slowly and ineffectively. Both of them had a chance to test their newly adopted emergency systems, which proved ill-conceived for such an emergency. They were also more concerned with long-term market developments than with the current crisis situation. Measures were implemented too late and did little to calm the market and stop the price rise. In addition, most member states did not implement the decisions. Interestingly, the most reasonable crisis-related measures were taken at an informal summit in Tokyo at the end of June. These, however, also remained only on paper. Unilateral action was the norm again.

While we can never be completely certain, this pattern suggests that coordinated emergency policy would be a very difficult task to accomplish, in the absence of strong leadership. This is the case at least in times of severe oil supply disruptions. Whether this assumption holds for the less critical cases will be examined in the next chapter.

Notes

1. Charles Mohr, "Carter Asks Strict Fuel Savings," *New York Times*, 19 April 1977.
2. Kapstein, *The Insecure Alliance*, 183.

3. Nikki R. Keddie and Yann Richard, *Modern Iran: Roots and Results of Revolution* (New Haven, CT: Yale University Press, 2003), 214–20; Milani, *The Making of Iran's Islamic Revolution*, 195.

4. Keddie and Richard, *Modern Iran*, 238.

5. Ibid.

6. U.S. Energy Information Administration, "Petroleum Imports by Country of Origin," http://www.eia.doe.gov/emeu/aer/txt/.html (accessed 17 December, 2010).

7. "BP Suspends Third Party Crude Oil Deliveries," *Oil & Gas Journal*, 3 September 1979.

8. Yergin, *The Prize*, 705.

9. "BP Suspends Third Party Crude Oil Deliveries."

10. Yergin, *The Prize*, 685.

11. U.S. Energy Information Administration, "OECD Countries and World Petroleum Demand, 1970–2008," http://www.eia.doe.gov/emeu/ipsr/t46.xls (accessed 15 March 2011).

12. Peter R. Odell, *Oil and World Power* (Harmondsworth, Middlesex: Penguin Books, 1986), 248.

13. Yergin, *The Prize*, 681.

14. Kapstein, *The Insecure Alliance*, 186.

15. "Iraqi Move Adds to OPEC Price Turmoil," *Oil & Gas Journal*, 18 June 1979.

16. "Aftermath of a Revolution," *Petroleum Economist*, April 1979.

17. Georges-Picot, *The Real Suez Crisis*.

18. U.S. Energy Information Administration, "Petroleum Imports by Country of Origin."

19. Lieber, "Energy, Economics and Security in Alliance Perspective," 151.

20. For a detailed account of conservation measures, see International Energy Agency, "Energy Policies and Programs of IEA Countries," (Paris: IEA, 1980). http://www.iea.org/publications/free_new_Desc.asp?PUBS_ID=1271 (accessed 15 March, 2011).

21. "What Oil Crisis?," *The Economist*, 11 August 1979.

22. "Control of Crude Oil Supplies," *Petroleum Economist*, September 1979.

23. "Germany," *Business Week*, 9 July 1979.

24. Robert J. Lieber, "Europe and America in the World Energy Crisis," *International Affairs (Royal Institute of International Affairs 1944-)* 55, no. 4 (1979): 539.

25. Lieber, "Energy, Economics and Security in Alliance Perspective," 152.

26. "News in Brief: United Kingdom," *Petroleum Economist*, October 1979.

27. David A. Deese and Joseph S. Nye, *Energy and security* (Cambridge, MA: Ballinger Pub. Co., 1981), 201.

28. Kapstein, *The Insecure Alliance*, 186.

29. Peter Goldman, "Carter's New Energy Plan," *Newsweek*, 2 April 1979; David Paully, "Over the Oil Barrel," *Newsweek*, 9 July 1979.

30. Ronald Koven, "Schlesinger: U.S. Losing Oil to Europe," *Washington Post*, 23 May 1979.

31. Ronald Koven, "IEA to Study European Oil Buying by US," *Washington Post*, 8 June 1979; Lieber, "Energy, Economics and Security in Alliance Perspective," 157.

32. Paulo Palango, "Canada Must Suffer Shortages Like Other Nations, Oil Man Says," *The Globe and Mail*, 16 February 1979; Anthony J. Parisi, "Attack on Oil Sharing Stirs Concern in Industry," *New York Times*, 17 February 1979.

33. Steven Pattner, "A New Distribution in World Oil Sales Perils U.S. Imports," *New York Times*, 15 March 1979.

34. Kapstein, *The Insecure Alliance*, 174.

35. Ibid.

36. Scott, *The History of the International Energy Agency.*

37. Ibid., 115.

38. James A. Phillips, *The Iranian Oil Crisis* (Washington, D.C.: Heritage Foundation, 1979).

39. Scott, *The History of the International Energy Agency.*

40. Ibid., 116.

41. Ibid., 117.

42. Article 17, Chapter 3 of the International Energy Program.

43. "Warnings on Oil Shortage at IEA Meeting," *Petroleum Economist*, June 1979.

44. Ibid.

45. A. C. Evans, "European Community Law and the Problem of Oil Shortages," *The International and Comparative Law Quarterly* 31, no. 1 (1982): 8.

46. Ibid.

47. Ibid.: 10.

48. "The European Council, Paris, 12-13 March 1979," http://aei.pitt.edu/1458/01/Paris_march_1979.pdf (accessed 16 December 2011).

49. "Energy; Not So Crude," *The Economist*, 8 December 1979.

50. Leonard Downie, "Britain to Reduce Oil Output," *Washington Post*, 4 January 1980; "Energy; Not So Crude."

51. "German Oil," *Washington Post*, 4 January 1980.

52. "Roy Jenkins, President of the Commission of the European Communities: Address to the European Parliament, 19 July 1979," http://aei.pitt.edu/11341/01/48-Jenkins-07-19-1979.pdf (accessed 13 January 2011).

Chapter 6
The Rest of the Cases

This chapter examines four smaller cases of supply disruptions. They are considered smaller for different reasons. All of them caused less damage to the world economy than the big 1970s crises. The physical shortage of oil, however, in some of the cases was comparable to the 1970s. As we shall see, there are a number of different reasons leading to this causal discrepancy. A close examination of the events surrounding each case, however, points to the conclusion that better international crisis coordination was the primary factor behind the smaller impact. For example, in 1951, Europeans lost more than half of their refined oil, as a result of the closure of the Abadan refinery. This was the second biggest postwar shortage after Suez. Its impact on European economies, however, was insignificant, mostly due to the assistance they received from overseas. The Six-Day War embargo caused difficulties for some consumers and not for others. The Iran-Iraq War led to a price increase of similar proportions to the 1978 shortage, however, it was much shorter-lived, because of market conditions at the time. The last case—the First Gulf War—led to the third biggest postwar oil shortage in absolute terms, removing 32.4% of European oil consumption from the market. This case needs to be examined in its historical context, as it happened in a very peculiar time of history—at the end of the Cold War.

Finally, the shortage caused by Hurricane Katrina in 2005, is briefly surveyed, being the most recent case of a supply disruption, and also the only post-1970 instance of IO policy coordination. Table 8 provides an overview of all small cases.

Table 8. Small crises

	Shortage in Europe (% of consumption)	US production (% of consumption)	International body
Abadan crisis	54	96	OEEC oil committee
Six-Day War	15	81	OECD oil committee
Iran-Iraq War	14.5	60	IEA
First Gulf War	32.4	52	IEA
Hurricane Katrina	0	33	IEA

The Abadan Crisis

In 1951 the world was still recovering from the bloodiest and most destructive conflict in human history—the Second World War. Western Europe was in the midst of its grand rebuilding project, sponsored by America. America was at the height of its power—the undisputed leader of the Free World. The United States produced 96% of the oil it consumed and had a locked-in capacity that could be quickly put in operation. The oil market was in the hands of the oil majors, which controlled the entire chain of operation—from exploration and production to downstream and marketing activities. Five of the seven oil majors were American. It was still twenty years before this configuration changed, however, the first signs of turmoil were already appearing. In May 1951, the Iranian government nationalized oil facilities in the country. A month later, oil shipments from the biggest refinery in the world completely halted. Abadan, at the time, was supplying 16% of OEEC refined oil and more than 25% of all refined products outside the Western hemisphere.[1] In a time of very low refinery capacity in Europe, and a war in Asia, this was going to be a major crisis. The situation, however, never spiraled out of control, mostly due to American companies' efforts in coordination with the administration. Large spare production and refinery capacity, and early-on emergency planning proved to be critical in this case. Two months after the Abadan closure, the European oil deficit had been overcome.

The Oil Shortage

The refinery and all oil extraction facilities in Iran were property of the Anglo-Iranian Oil Company (AIOC)—the future British Petroleum—operating under a concession agreement with the Iranian government. AIOC had the right to exploit and sell Iranian oil for sixty years, starting in 1933.[2] With the rise of the nationalist movement in Iran, and in particular the National Front, headed by Mohammed Mossadeq, the terms of the agreement came under attack. Negotiations to revise the concession started in the late 1940s. An initial 50/50 deal, proposed by the Iranian government was not accepted by the company. In the protracted bargaining that followed, compromises were difficult to reach due to the increasingly nationalistic atmosphere in Iran, and the intransigence of the British government, which at the time was AIOC's majority owner. As a result, in March 1951, the Iranians decided to nationalize the company's assets. Two months later, AIOC imposed an embargo on all oil shipments from the refinery, closing all facilities. The last British personnel left Abadan in October.

The Abadan refinery had a very important place in the European oil supply system. While there was plenty of crude oil to be found outside of the Middle East, refined products were harder to replace. In 1951 Western Europe had a very low refinery capacity, relying primarily on imports from Iran or the Western hemisphere. The shortage of aviation fuel was particularly severe. At times, there was a deficit of 20,000 b/d of aviation gasoline or half of daily European consumption.[3] This posed an enormous risk for the aviation industry, and grounding of flights appeared a real possibility. A major crisis was averted by the swift and timely reaction of the American administration.

The United States

The American reaction was guided primarily by Cold War considerations. The impact of the oil shortage on European economies was a major concern. In several memoranda, estimates and letters, the State Department gave assurances that Europe would be supported in the event of an oil supply disruption, since "any faltering in free world unity" might tempt the Soviets to move into countries like Iran, Yugoslavia and Germany.[4] It was also understood that "Soviet control of Iran…would deprive the West of a vast oil supply."[5] This would significantly delay European recovery and create severe financial difficulties for Great Britain. Since the Abadan refinery was the largest in the world, its closure would have implications for the entire oil industry. There was also a risk of a communist takeover within Iran, especially since the communist Tudeh party had been increasingly gaining popular support.

In this situation, urgent action was necessary. Already in June, the government put in motion the necessary machinery. This swiftness is remarkable given that the United States, at the time, was engaged in war on the Korean peninsula. On the other hand, because of the war effort, there was already an oil emergency institution in place—the Petroleum Administration for Defense—

which expedited the process. It is within this body that the Foreign Petroleum Supply Committee was formed. It was established under a "voluntary agreement" between industry and the government, which was to advise the administration on foreign oil supply operations. In cases of emergencies outside the US, the FPSC was to prepare plans "to prevent, eliminate or alleviate shortages of petroleum supplies, which threaten to affect adversely the defense mobilization interests or programs of the United States."[6] This committee would also play a critical role in the oil crises of 1956 and 1967.

In 1951, the FPSC presented its "Plan of Action No. 1" as early as July. According to the Plan, crude oil and refinery production had to be increased in a number of countries, including the United States, while in the meantime, American stocks of products were to be shipped to European allies.[7] This uneasy task was to be coordinated by the oil majors, which were exempted from antitrust regulations. This was a critical decision that was later put under investigation. At the time of the crisis, however, it was highly necessary. An operation of this magnitude needed the exchange of information among all companies involved, and among companies and state institutions, which would have been impossible under anti-trust regulations.

Similar to the Suez crisis, oil majors and governments coordinated emergency action on the other side of the Atlantic, as well. British oil companies met with American counterparts in the British Oil Supply Advisory Committee in London, formed within the Ministry of Fuel and Power.[8] The companies coordinated crude production increases in Saudi Arabia, Kuwait and Texas and refinery output expansion in the United States.[9] In August, 200,000 b/d of refined products were shipped to Europe from American ports.[10] In addition, Middle East oil destined to the United States was diverted toward Europe. Despite its complexity, the entire operation ran very smoothly and two months after the closure of Abadan, the oil deficit experienced by "friendly foreign nations" had been met.[11] The secretary-general of OEEC declared in August that Europe was no longer "alarmed" by the loss of Iranian oil.[12] The emergency measures taken by the companies proved, indeed, crucial in preventing a major oil shortage and a scramble for resources in Europe. Consequently, the crisis received little to no attention in the French and German press at the time. The French *Revue Economique* mentioned the oil shortage only once, in relation to the balance of payments difficulties of the British government.[13] The OEEC oil committee did not devote much attention to the crisis, either.

The OEEC Oil Committee

The first years of the oil committee were marked by efforts to coordinate refinery expansion in Europe. This was an important task, which was successfully accomplished by the end of the 1950s. No oil emergency powers were vested within the oil committee. At the time of its establishment in 1948, oil was a relatively minor energy resource in Europe, compared to coal, and war recovery was the main focus. In general, the main purpose of the Organization for European Economic Cooperation was to coordinate American aid to Europe. Its oil committee, which was the only international energy body at the time of Abadan, was also formed to serve the needs of the disbursement of American aid. There were no emergency procedures in place, and dealing with distribution and coordination problems was not envisaged. The committee's only involvement in the Abadan shortage was to share information on the allocation plans being developed within the United States and UK emergency bodies.

The committee's lack of emergency powers and its preoccupation with other issues could partially explain its minor involvement in the crisis. However, there is another candidate explanation. It became clear early on that American companies would be able to cover the shortage. In 1951, the United States was still heavily involved in European reconstruction. It was the mainstay of the Western alliance, and the only country therein capable, economically or militarily, of exerting influence. There were shared expectations that the United States would come forward in case Europeans experienced an emergency of any kind. The fast reaction of the American government confirmed these expectations and allayed European fears. In December, the chairman of the oil committee declared that "this happy cooperation on both sides of the Atlantic enabled the situation to be met without too many difficulties."[14]

The Abadan Crisis: Conclusion

The 1951 shortage was successfully overcome within a very short period of time, even though turmoil in Iran continued until 1954. The government of Mossadeq was overthrown in the fall of 1953, and a year later a consortium of eight international oil companies was formed. Anglo-Iranian was the main shareholder with 40%. The recently formed National Iranian Oil Company became the official owner of oil facilities in the country. The oil majors, however, exploited and sold the oil.

The United States' swift reaction to the crisis is the main factor behind its fast resolution. American extraction and refinery output could easily be increased at the time—the United States was still the largest oil producer and exporter in the world, producing 90% of its local consumption. This favorable situation combined with the reality of low oil consumption in Europe—oil accounted for only 18% of the energy mix. While this number was much higher in 1967, United States' power in oil was still strong.

The Six-Day War

The 1967 oil shortage was a consequence of one of the most defining moments in the turbulent history of the Middle-East. The Six-Day War had a transformative impact on the region, and marked the beginning of a new phase of Arab-Israeli relations. In terms of energy security, the war led to one of the most serious oil shortages in OECD Europe. For a short period of time, all sources of Middle Eastern oil, with the exception of Iran, were interrupted. This represented more than 80% of Europe's oil—a shortage similar to the one in 1973. Yet, this time, the crisis went almost unnoticed, and its consequences were relatively minor. As in 1951 and 1956, the successful crisis management is to be attributed mostly to American leadership. Several factors deserve attention. First, the United States administration was able to organize quickly, sending a message of reassurance across the Atlantic. An emergency plan of action was approved as early as 29 June. Second, the available spare production capacity was quickly brought on line, and shipped to European allies. At the same time, mothballed tankers were put into operation, while others—diverted from their regular courses—toward European ports. In this respect, there were no major differences between the 1950s and the 1967 oil crises. A third factor, however, also played an important role in 1967—American coercive power. US leadership was instrumental in bringing allies to coordinate efforts within the OECD oil committee.

The Oil Embargo

The days leading to the conflict were a typical example of a crisis escalation, in which both perceptions and facts on the ground played an important role. Regular skirmishes on the Israeli-Syrian border and Syrian-sponsored Fatah terrorist attacks on Israeli territory were the building stones. Then in May 1967, the UN peacekeeping force, stationed on the Sinai Peninsula since the Suez crisis, left at the request of Egyptian president Nasser. His ambitions to unite the Arab world under his leadership, and his belief in the superiority of Arab forces, shaped his policies in the moths before the war. Shortly after the UN withdrawal, Egyptian forces moved into the peninsula up to the Israeli border, and Nasser declared the straits of Tiran closed for Israeli ships. For Israel, this was considered a *casus belli* and on June 5, 1967, Israeli air forces attacked Egypt, destroying almost completely its air force. After five days of intensive fighting, which destroyed most of Egypt's, Syria's and Jordan's military capabilities, a cease-fire was signed.[15] As a result of the war, Israel acquired the Golan Heights, the Western Bank together with East Jerusalem, and the Sinai Peninsula.[16]

Incidentally, a day before hostilities started, eleven Arab oil producers gathered in Damascus to consider embargoing countries supporting Israel. An oil embargo was declared, and the Suez Canal was closed for all shipping. Some oil producers (Saudi Arabia) halted all oil exports. Others, like Algeria, continued exporting, applying restrictions only to the United States and UK. Toward the end of June, West Germany was also included in the embargoed states, even though the German government had declared neutrality at the very beginning of the conflict.[17] Its prior military support to Israel must have played a role in the Arabs' decision. As in 1973, however, Arab producers could not achieve unity and the embargo broke shortly after it was announced. As early as June 11, Saudi Arabia resumed exports to all destinations with the exception of the United

States and Britain. On September 1, 1967, Arab leaders, meeting in Khartoum, decided to put an end to the boycott. By that time, it had inflicted more damage on producers' than on consumers' economies—a fact admitted two months earlier by the Saudi Arabian oil minister.[18]

Without a doubt, the Arab embargo did have some impact on consumers. The United Kingdom was the most affected country, as it was fully embargoed. It was the only country to adopt a rationing plan. Not surprisingly, the UK representative in the oil committee firmly stood behind the US proposal for emergency sharing. Unfortunately, as will be discussed later, most other countries did not feel obliged to help a fellow member. France was not among the embargoed states, and was less affected by the closure of the Suez Canal and the pipelines, as it imported 29% of its oil from Algeria.

Germany was in a similar situation. It was part of the "enemies" list, however, it got 36.3% of its oil from Libya, which was not affected by the closure of the Canal and pipelines. In addition, it was added to the embargoed countries later, and stopped receiving oil supplies at the end of June—shortly after the United States enacted its emergency plan. In July, the first tankers loaded with American oil arrived in Wilhelmshaven.[19]

The United States

In 1967, the United States was still satisfying 85% of its oil needs with domestic production. The dependence on Middle East oil was yet to come. Moreover, the unused production capacity, which could be brought in line, was still relatively high. Interestingly, only six years later, American oil production would already be in its post-peak period with no spare capacity.

The US actions were again prompted by Cold War considerations. Europe's reliance on oil had increased dramatically during the 1960s. Oil was now 50% of the energy mix, as compared to 30% in 1956.[20] 83.8% of crude oil came from Middle East sources (including North Africa).[21] Crude oil throughput had increased 228%.[22] Depriving Europe of its lifeline could have devastating consequences for both European and American security. Consequently, Western Europe's oil provision was at the top of US priorities during the crisis. The other one being oil supplies for the war in Vietnam. US combatants in Vietnam received about 200,000 b/d, which came entirely from the Middle East.[23] Fortunately, Arab producers did not link Vietnam with the US presence there, and continued shipping oil to the country.

The American reaction was fast and swift. It followed a pattern similar to 1956. On June 8, a meeting of twenty-one oil companies with government officials took place in the Department of Interior, reviving the Foreign Petroleum Supply Committee—so instrumental during the Abadan and Suez crises.[24] The committee was instructed to prepare plans for emergency supplies to affected countries. The emergency program, including antitrust exemption, was approved on June 29 and American oil started flowing to Europe shortly after. A separate body—Emergency Petroleum Supply Committee—was set up to implement the plan.[25] As in 1956, American companies took two main actions: increasing output in oil-producing states, and rerouting or putting into operation mothballed tankers. From May to August 1967, Texas and Louisiana increased production with 1 million barrels a day. Venezuela and Iran also increased exports. In total, during June and July, American companies shipped 8 million barrels to Europe. Similarly to the Suez crisis, the shortage was mostly of tankers and not crude, which could be found in non-Middle East producing countries. As admitted by the OECD report, "the problem of re-organizing sea transport was not easy."[26]

The OEEC Oil Committee

The Six-Day War embargo provided the first test of the oil committee emergency preparedness. In 1951 and 1956, the committee did not have emergency functions. After Suez, various energy security measures were considered, and a scheme was adopted in 1962.[27] Building oil reserves was one of the first and most discussed measures. There was even a proposal to create a common stock pool, held by an international body, with member-states as shareholders. This idea was not accepted, and instead, the committee agreed on national stockpiles of minimum 60 days of consumption. At the time of the crisis, states had stocks equal on the average to 71 days of consumption (see table 9).[28] While the stock concept is still considered one of the backbones of oil energy security, it has not proven its usefulness yet. For stocks to be effective, their drawdown and accumulation need to be implemented in a concerted manner. And coordinated action, especially in an area so vital for states' survival, has proven difficult to achieve. The availability of stocks does not provide a security buffer to the same degree as the availability of spare capacity in friendly hands and the ability or willingness to use it. Stocks were at their highest levels in 1978, and yet this did not prevent scrambling. In general, states have been more inclined to fill in stocks in times of crises, thus exacerbating the shortage.

Table 9. Oil stocks, 1967

	Oil stocks (days of consumption)
France	81
Germany	68
United Kingdom	86

Source: Data from High Level Group of the Special Committee for Oil, "Draft Preliminary Report on the 1967 Oil Emergency," (Paris: OECD, 1968).

An important element of the 1962 emergency system was the industry body. If an oil shortage were to happen, an international industry advisory board, similar to the 1956 OPEG, would to be set up. The Board had to assist the committee with the organization and implementation of an eventual apportionment of supplies.[29]

Even more important, however, was the recognition of the American role. In case of an emergency, the United States government was to mobilize its committee system, and in particular the MEEC, which would then participate in oil committee deliberations in Paris. OPEG—so critical during Suez—was also to be activated, and was to participate in the MEEC work. Member-states were aware of the importance of America's capacity to intervene. This awareness, together with the successful Suez experience, gave them the confidence that they could easily manage another oil crisis. Indeed, in terms of emergency cooperation, the 1967 shortage provided another positive experience.

The role of the United States—the administration and companies together—is indisputable. This transpires from the records of oil committee meetings, where there was an opposition to the American-sponsored declaration of emergency. As it will happen later, in 1973, some European countries preferred to tackle the shortage unilaterally, not to antagonize Arab producers. According to Germany and France, countries had enough resources to handle a crisis of that magnitude on their own. After all, stock levels were at their highest, and the oil shortage was nothing like the 1956 crisis of Suez. These attitudes became clear at a meeting of the oil committee on June 12. Unlike in 1973, however, this time the United States had the resources to coerce Europeans into a common solution. The United States representative, baffled by the inaction of the committee, warned that US companies would not be authorized to share information if a common declaration of emergency were not adopted. This threat proved to be sufficient and a motion, stating that the threat of an emergency existed, was adopted. France, Germany and Turkey abstained.[30] The oil committee's resolution paved the way for companies' concerted action. An International Industry Advisory Board was put in place, to consult the committee on oil supplies' availability and movements. Working together with the EPSC in the United States, the Board devised plans for increased production and tanker availability and successfully countered the shortage. As the crisis quickly abated, oil allocations were not necessary this time. An additional advantage was the increased tanker tonnage after 1956, leading to an oversupply of tankers before the embargo. As recognized by the oil committee: "owing to joint action by member governments and the international oil industry, unlike in 1956/1957, no allocation plans had to be applied."[31] By September, the oil market was going back to normal. The IIAB was put on standby in January 1968.

The Six-Day War: Conclusion

In comparison with the other oil shortages, 1967 could be ranked as the most serious one among the second category crises. It created a shortage of similar magnitude as Suez, however, its consequences were much less severe. To start with, as a consequence of the 1956 crisis, Europeans had diversified their oil supplies, now receiving 7% from the Soviet Union and Eastern Europe.[32] Also as a result of Suez, when tanker shortage proved to be the biggest problem, the tanker fleet had dramatically changed. In 1967, tankers were much bigger than in the1950s, which made the route around the Cape of Good Hope not so onerous anymore. In addition, the Suez Canal had lost its significance as the main oil transport artery. As assessed by a *New York Times* editorial, the effect of the Canal closure was "more a dislocation than a disruption."[33] Yet, some European countries got deprived of almost 80% of their oil. This shortage was covered, as in the two previous crises, with increased US production and tanker capacity. Timely American emergency preparations and leadership within the OECD oil committee were very critical. It is difficult to predict how the crisis would have evolved without them. However, the high level of market interconnectedness would have made it difficult for individual countries to secure supplies without hurting the supplies of others. More importantly, states would have to possess high tanker capacity to cover for the loss of pipelines and the closure of the Suez Canal. Only the United States could bring up additional tankers at the time. In addition, American companies had a controlling share in oil production in most non-Arab countries where output could be increased (with the exception of Iran). All this made American leadership necessary.

The Iran-Iraq War

The Iraqi invasion of Iran happened at a critical moment in Middle Eastern and oil history. Only a year ago, the world was in the midst of one of the biggest oil shortages ever, precipitated by revolution in Iran. Oil supplies were down 10% (as much as 14% in some countries) and prices rose 150%. The whole year 1979 was marked by political turmoil in the Middle East and other producing regions, which directly affected the oil market. A terrorist attack in Saudi Arabia, oil production nationalization in Nigeria, and the continuing Iranian hostage crisis all contributed to the atmosphere of uncertainty in the market. Nevertheless, by early 1980, the market was on the road to recovery, and in the summer, there was an excess of 3 mbd of oil. Then crisis struck again.

On 22 September Iraqi troops invaded Iran. There had been border skirmishes between the two countries for more than a month before that. The official pretext for invasion was a long-standing territorial dispute over control of the Shatt-al-Arab waterway—an important channel for both countries' exports. There were also a number of broader underlying concerns. Among them, fear of Khomeini's revolutionary zeal spreading throughout the Middle East and undermining Iraq's influence in the region, was an important factor. Personal dislike between Saddam and Khomeini, and deep-rooted historic animosities between the two countries, based on ethnicity, religion and ideology, also played a role.

The onset of hostilities between Iran and Iraq led to pipelines and refineries closures. For a short period of time, almost the entire production of the two belligerents was removed from the market. And as in the two previous crises, in 1973 and 1978, market behavior was dominated by panic. American spare production capacity was long gone, and while inventories were high, uncertainty about the duration of the war and its impact on the region remained. The Carter administration, bogged down in economic recession, the Iranian hostage crisis and a reelection campaign, was not capable and/or willing to take the lead. And the International Energy Agency did not have the capacity to replace it. The lack of crisis leadership led, for a third time in a decade, to scrambling for supplies and high prices.

The Oil Crisis

Iraqi forces attacked oil-producing facilities in Iran, including the big Abadan refinery. During the Iranian counterattack in November, Iraqi oil installations were destroyed and Iraq's access to the Gulf was closed off. This removed from the market 3.3 million b/d or 22% of OECD Europe consumption, for a period of three months.[34] In January, limited amounts of oil started flowing and the market slowly returned to normal.

The immediate reaction of the market was again, and for a third time in seven years, based on fear. Spot prices quickly rose and Arab crude reached its highest price ever—$42 a barrel.[35] The huge inventories built by companies and governments were suddenly not considered sufficient and panic buying started again. OPEC ministers did not wait long to profit from the situation and raised prices to $36 a barrel in December.[36]

This time, however, markets were much quicker to stabilize. The Saudis were able to increase output with 900,000 b/d in just a couple of days, covering almost one-third of the lost oil from Iran and Iraq. Other producers, including Britain and Norway, also increased production. Combined with fast falling consumption, this led to an easing of the market situation as early as December. In the beginning of 1981, prices were already declining. The Third Oil Shock, predicted by some, did not materialize.[37]

The United States

The US administration did not take any action, this time, to quell allies' fears. There was no production capacity to be increased, and no coordination with the oil companies took place. In the course of the United Nations' opening session at the end of September, American diplomats discussed with their European counterparts the dispatch of an international naval force, to protect oil shipping through the Strait of Hormuz.[38] This idea did not come to fruition possibly because it became clear that Iran was committed to keep the strait open. In any case, no initiative was taken with regard to the oil market. The Carter administration adopted the view that there were enough oil inventories to make up for the loss and therefore no emergency existed.[39] Stocks, in fact, were not much higher than in 1967, when the American government got involved in the crisis early on. However, domestic and international political conditions might have prevented Carter from taking a firmer stance this time. By 1980, his presidency had become very unpopular, making it difficult to forge consensus or introduce policy on any issue. At the time of the crisis, he had been involved in the Iranian hostage debacle for almost a year. His efforts to free the hostages had been unsuccessful. His entire presidency was marked by rising inflation and unemployment—an economic crisis created partially by the First Oil Shock, but also by some of his policies. To top it all, at the time of the oil crisis, he was busy with campaigning for reelection. Thus, the oil shortage never made it to his priorities list. The oil companies and, eventually, the International Energy Agency would have to lead the way. Without antitrust exemption, however, American companies would not share information and work together to fill in the gaps produced by the shortage. Let us recall that their coordination had been essential in overcoming the shortages of the 1950s and 1960s. Times have changed, however, and companies were on their own.

The US inability to gear events also reflected in the IEA deliberations. There, the American representatives suggested the introduction of individual import ceilings—an old Carter initiative.[40] The proposal never went through, however, as some countries feared it would lead to gasoline rationing. In general, the IEA efforts during the crisis were limited to recommendations, which, most of the time fell on deaf ears.

The International Energy Agency

Soon after its first test during the Iranian revolution, the International Energy Agency was given another opportunity to demonstrate its utility. This time, reaction was faster, but not very effective. To be sure, there had been a learning process between the two crises. The Governing Board met several times during 1980 to evaluate its previous actions and to consider possible amendments to its emergency policies. These deliberations led to a decision to increase stockpiles to 90 days of net oil imports. A concerted stock draw came to be considered as the most appropriate action in the case of a shortfall. Stockpiles served two purposes: buffering the initial shock and deterring producers from using oil as a weapon.[41] Allocation of supplies, as envisaged in the International Energy Program, was increasingly seen as a nonworkable proposition. The increased reliance on stocks still remains the backbone of the IEA crisis response system.

In 1980 the board also began discussions on eventual emergency measures for shortfalls of less than 7%. While concrete measures were not adopted until 1984, the energy ministers agreed to meet quickly for consultations if a shortage of any scale appeared imminent. Their reaction in September 1980 was faster than in 1978. Member-states met as early as 1 October and agreed to refrain from "abnormal" spot market purchases and to begin consultations with the oil companies.[42] Immediate release of stocks was encouraged. These decisions were reconfirmed at the next meeting in December. Members, in effect, were called on to refrain from building their stocks at the moment, and use them instead. However, a stockdraw—coordinated or not—never took place. On the contrary, some countries engaged again in competitive bidding and stock filling. For example, France and Belgium were reported to be excessively buying refined products on the Rotterdam spot market in late October, that is, after the IEA decisions were taken.[43] Undoubtedly, this uncoordinated crisis behavior was the main reason behind the price rise.

In addition, in two important cases—Japan and Turkey—the Agency failed to secure coordination. Japan continued bidding for supplies and pursuing bilateral deals at the height of the crisis. The government signed a supply contract with Kuwait, which included a premium of $5 per barrel.[44] Inevitably, this contributed to the overall price increase. Some developing countries—which were the hardest hit by the shortage—also resorted to bilateral deals. Brazil, for example, which imported 40% of its oil from Iraq, negotiated emergency supplies with Indonesia, Venezuela, Saudi Arabia and other producers. These countries were also bidding on the spot market to ensure supplies.[45] The pressure was partially relieved by Saudi Arabia, which released a list of priority destinations for its oil in November. Among these countries were the heaviest bidders in Rotterdam—India, Brazil, France, Japan and Turkey.[46]

The Turkish case is especially interesting. In a sort of a replay of the 1979 situation with Sweden and Denmark, the IEA proved incapable of organizing assistance to one of its members. Turkey was the hardest hit by the oil shortage, as it imported about 60% of its oil from Iran and Iraq. The country was also short of foreign reserves, and low on oil stocks—factors, which significantly exacerbated the situation. As a last resort, the government asked the IEA for help. The ensuing negotiations between the IEA secretariat, oil companies and the United States did not lead to an agreement. Turkey was offered supplies by several oil companies, however, it could not afford to pay the high-priced oil. In mid-December, oil supply in Turkey reached its bottom, with widespread shortages around the country and stocks available for less than five days.[47] The situation was eased only by the partial resumption of oil flows from Iraq.

The EEC

The European Community did not take any particular action this time. At the end of November, the Nine took a decision to use their stocks to prevent a further price rise. There was one condition, however—that other industrialized countries do the same. This immediately transferred responsibility to the IEA, where all developed nations participate. As market conditions improved in December, and fears of a Third Oil Shock abated, no further action was taken.

The absence of a concerted EEC reaction could be attributed to differing member-states relations and attitudes toward the two warring countries. These also brought different patterns of oil imports and trade relations, in general. For example, France imported 34% of its oil from Iraq, while Germany was somewhat less dependent on any of the two countries. Being one of the hardest hit among Western Europeans, France was also among the hardest bidders in Rotterdam. Differences of this sort have prevented common European action in the past, as well. They are indicative of the difficulties of building a common foreign and security policy, which still persist today.

The oil shortage provoked by the Iran-Iraq War lasted only three months. The war, however, continued until 1988. Iraq's only oil outlet for most of the war duration was a pipeline through Turkey with a 750,000 b/d capacity.[48] Iran's exports averaged only 2–2.5 mbd for the period 1980–1984. The reason for this situation's low impact on the oil market is not to be found in the US leadership or the IEA mediation. It was rather due to the prolonged effects of the two 1970s oil shocks, both in terms of economic recession and conservation measures. World oil consumption reached its lowest point in 1983. It started to increase again afterward as the economic situation improved, but remained well below its 1979 peak until 1989 (see figure 5). Producers had to considerably reduce output and in the second half of 1982, Saudi Arabia's production fell from 10 mbd to 6 mbd. By mid-1985, it had declined to 2.3 mbd.[49] A congressional report of 1987 noted that even if Persian Gulf exports were cut off, the impact on supply and prices would be minimal.[50] In the words of Peter Odell, "too much oil was chasing too few markets."[51]

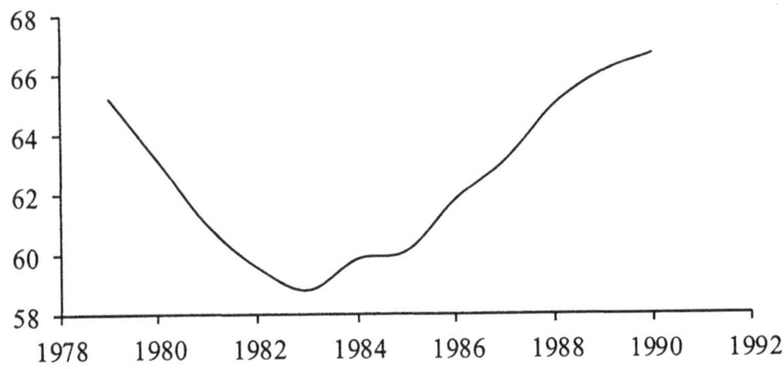

Figure 5. World Oil Consumption (million barrels a day)
Source: Data from U.S. Energy Information Administration, "World Petroleum Consumption, 1960-2007," http://www.eia.doe.gov/emeu/aer/txt/ptb1110.html (date accessed 17 December 2011)

The Iran-Iraq War: Conclusion

The 1980 crisis was of similar magnitude as the one a year earlier. However, it lasted much shorter and did not have such a severe impact. This has prompted some analysts to declare the case a relative success for the IEA.[52] Others, however, disagree. Yergin, for example, argued that "The IEA's efforts to encourage inventory use, instead of panic buying, were not meeting with great success."[53] The fact that some panic spot buying took place and oil prices increased significantly lend some evidence in support of the latter argument. The shorter duration of this crisis behavior, and by extension of the crisis as a whole, might be due to two other factors. First, oil market conditions were more favorable this time. Conservation measures introduced as a result of the 1979 shortage, and the recession it caused, led to a considerably lower consumption. Average world demand in 1980 was 2 mbd lower than its peak in 1979, and in 1981—another 2 mbd lower. Again as a result of the previous oil shortage, commercial stocks were at one of their highest levels. Companies were storing oil in supertankers rather than selling it at a loss on the market. The market was so oversupplied that OPEC took a decision in September 1980 to cut back production by 10%.[54] This left producers with a large surplus capacity, which might have played a role in reducing market anxiety.

Second, some sources suggest that the initial surge in bidding and prices was partially caused by the fear that military action could close off the Strait of Hormuz, through which most of the exported Persian Gulf oil—or some 12.70 mbd—passed.[55] A shortage of that magnitude would deplete the world's spare capacity and stored oil in a very short time. This concern was also reflected in the initial official statements of the US administration and the EEC regarding the war, both of which stressed the importance of the Strait.[56] As the threat to Hormuz eased later, this fear subsided, and the supply situation returned to normal. That is, given the lax supply conditions, the war shortage in itself was not sufficient to stir the market. Very quickly it became obvious that the interruption of Iranian and Iraqi oil supplies could be absorbed without much disturbance. Moreover, as early as December, both countries partially resumed production, which additionally calmed market anxiety.

The First Gulf War

On August 2, 1990, Iraq invaded Kuwait. Not meeting much resistance, the Iraqi army took over the small country in two days. The war reduced oil supplies 7% of world consumption. Favorable market conditions, including high world spare capacity, contributed to the milder effect of the crisis. However, mostly due to uncertainty, market participants engaged in panic buying, which brought a price increase of 118% in October. The United States, unable to forge a consensus within the International Energy Agency, enacted a successful domestic test stockdraw. The IEA agreed on a coordinated stockdraw in January 1991, which was never applied by member-states. In any case, this decision came late, after lost production had been offset and the market had returned to a condition of oversupply.

The reasons for Iraq's aggression have been discussed at length elsewhere.[57] A few notable points deserve mentioning here. The immediate reason, according to Saddam, was Kuwait's overproduction of oil, which brought down prices. Iraq, at the time, was facing severe financial difficulties, mostly as a result of the 8-year war with Iran. Kuwait's oil reserves would have brought a significant increase in Iraq's oil income, as well as a cancellation of its enormous war debt. The Rumeila oil field—stretching across the border between the two countries—was also part of the equation. Saddam claimed the field belonged to Iraq, and even if he were to accept the status quo, he believed the Kuwaitis were extracting more than their share. There were deeper historic and ideological reasons, as well. Saddam's Baath party ideology of pan-Arabism and his aspirations to unite all Arab lands were the underlying motives of the invasion. His ambitions grew even bigger after the Iran-Iraq conflict, which saw him strengthening his position in the region. In addition, Saddam had always considered Kuwait as part of its southern province of Basra. During the Ottoman period, the present Kuwaiti territory was part of the Basra Ottoman province, most of which later became part of Iraq.

Saddam Hussein declared the annexation of Kuwait on August 8. A frantic diplomatic activity ensued, involving the United Nations, the United States, the European Community, separate European governments, and a number of Arab nations. After several negotiation rounds between the UN secretary-general and the Iraqi foreign minister, an economic and military embargo, and 12 UN resolutions, Saddam appeared determined to stay in Kuwait. The UN ultimatum expired on January 16, 1991. Coalition air forces, led by the United States, attacked on the following day. A ground assault followed in February and Iraqi troops started withdrawing at the end of the month. A cease-fire was declared on March 3, 1991.

The Oil Shortage

As a result of the Gulf War, 4.8 mbd were removed from the market.[58] Iraq and Kuwait accounted for about 7% of world oil production and 11% of EEC oil imports.[59] A disruption of this magnitude would have been catastrophic in a world of very low or no spare capacity. The 1973 experience is a case in point. Excess capacity in 1990, however, was about 3.8 mbd equaling 7% of current world consumption.[60] More than two million barrels a day of this surplus were located in Saudi Arabia, which was able to put them on the market in a very short period of time. Venezuela and the United Arab Emirates expanded production with 500, 000 b/d each and Indonesia—with 250,000 b/d.[61] Increases were also reported in Oman, Malaysia and Mexico, and later in the year—in Indonesia.[62] By November, Saudi Arabia had increased production with 2.9 mbd over July levels.[63] According to the IEA Oil Monthly Report from January 1991, world oil supply in December 1990 was 53.9 mbd, while consumption equaled 53.7 mbd.

There were two more reasons not to panic and scramble in 1990. In the first place, demand for oil was currently low, as a result of the downward economic cycle of the late 1980s. The *New York Times* discussed a 5% lower consumption of oil products as a result of the recession.[64] This also allowed Saudi Arabia and Iran to store large quantities of oil in tankers close to major consumer countries.[65] On a structural level, oil had become a much lower percentage of the energy mix of consuming countries. In France, for example, it represented 40% of energy usage, as compared to 69% in 1973. 80% of electricity in France in 1990 was produced by nuclear power. This, apparently, was an important factor in shaping public attitudes toward the crisis. According to Heisbourg, there was a widespread belief that France's nuclear energy gave it "an element of security," even though France imported a substantial amount of Iraqi oil and had major financial and industrial interests in the country.[66]

In the second place, as a result of the low oil prices of the 1980s, most governments had accumulated strategic reserves for 98 days—more than the IEA requirement of 90 days. Germany, Japan and the United States were the champions with stocks equal to 128, 110 and 100 days of consumption, respectively. France and the UK had stocks for about 80 days.[67] This discrepancy might explain the initial reluctance of Germany to participate in a coordinated drawdown. Oil companies had stocks for 70 days, which was 15 days above the requirement.[68]

Important market innovations, introduced in the 1980s, also contributed to alleviating shortage fears. At the time of the previous oil shortage, the market was dominated by long-term contracts. In 1990, the price of oil was set on commodities' exchanges, making it much more efficient and flexible. In 1983, oil futures were introduced on the New York Mercantile Exchange. This allowed for hedging against oil price swings, introducing a level of predictability. All these factors helped ease the fears of market participants and the oil price in 1990 rose much less, in comparison with 1973 (400%) and 1979 (150%). If all the above factors are added to the equation, according to an analysis in *The Economist*, the price of oil in 1990 had to reach $85 per barrel, to inflict the same economic damage as in 1979.[69]

Despite the favorable market conditions, and the willingness and ability of producers to quickly increase output, prices rose dramatically. In October the price was up 118% over pre-crisis levels. Similar to 1980, the price increase was mostly psychological, triggered by speculations about the duration of the conflict and the threat it posed to Saudi Arabia. The Saudis held 25.5% of world oil reserves. Taken together with Iraqi and Kuwait reserves, which at the time accounted for 20%, Saddam could have taken control over almost half of the world's oil. This was a serious threat, which market participants took into account. Although Saddam declared at the beginning of the conflict he did not intend to attack the Saudi Kingdom, there was considerable concern among Western allies as to his true intentions. As a result, without any coordination, some of them rushed to secure supplies. Panic buying was the norm again, bringing the price up to $35 for Brent crude at the end of September (see figure 6).[70] Stocks in Europe increased with about 225,000 b/d in the fourth quarter of 1990, suggesting the counterproductive pattern of the 1970s crises—filling up stocks in times of shortage, and using them in times of normalcy.

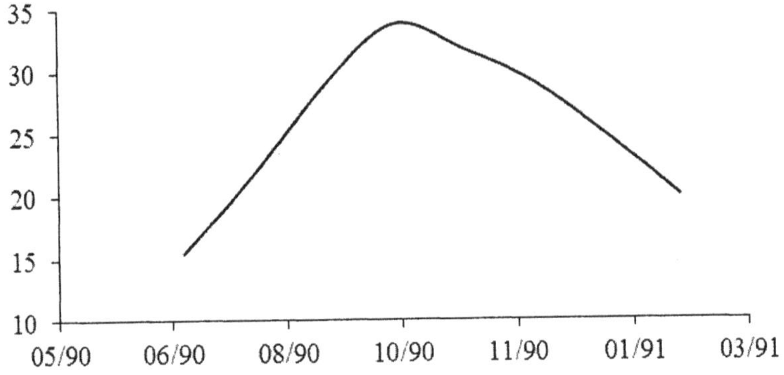

Figure 6. Average price of crude oil imported into IEA countries ($/bbl)
Source: Data from International Energy Agency. *Monthly Oil Market Report: April.* Paris: IEA, 1991

The United States and the International Community

The international response to the invasion was fast, united and unconditional. Iraq's aggression produced an unprecedented degree of international coordination—a feat that has not been repeated since then. It gave hope for a different post–Cold War future, one of cooperation without conflict. Between August 1990 and April 1991, the Security Council adopted 15 resolutions on the Gulf War, most of them with unanimity. Sanctions were imposed both by the Security Council and the European Union, while the Soviet Union placed an embargo on arms sales to Iraq—a critical move, since the Soviets were close Iraqi allies during the Cold War. The Arab League condemned the invasion with unanimity.[71]

The European Community initially achieved a relatively high level of accord, as well—a success for its dismal common foreign and security policy record. Divergences appeared later, with France taking a diplomatic initiative on its own, without prior consultations. Participating in the coalition effort was also a source of disagreements. At the end, almost all EC members took part in the coalition in one way or another. Even Germany, limited by its Constitutional ban to send troops abroad, provided assistance by contributing to the NATO operation in Turkey and sending minesweepers to the Gulf after the conflict.

The United States was, without a doubt, the most critical actor in organizing the diplomatic and military effort. As Dilip Hiro notes, by the end of August, "the crisis has shaped up as a confrontation between Iraq and America, with each side acting as a mirror image of the other."[72] On the oil front, however, America was not able to master resources and to provide leadership. The American spare capacity was long gone. Its production had declined with two million barrels per day in only four years—between 1986 and 1990.[73] Oil imports were at their highest levels providing almost 50% of consumption. In short, America could not bring others to cooperate as it has done in the past. Its unilateral stock drawdown, however, helped ease the pressure off the market and contributed to the price decrease in November.

Despite the relatively favorable market conditions at the time, there was awareness within the administration of a possible oil shortage as a result of the war. The US representative in IEA suggested a coordinated stockdraw as early as September. Such a release would have eased market fears, and taken pressure off prices. The idea was not accepted by the Board, opposition coming mainly from Germany and Japan. According to testimonials, the US government was ready to participate in a release, as long as other consumers contributed, as well.[74] Not being able to forge a cooperative solution within the energy body, the United States decided to act on its own. At the end of September, the government took a decision to test-release crude from its reserves.[75] Between September and December, 3.9 million barrels were released and 26 separate shipments of crude were made from three of the six storage locations. This amount was insignificant compared to the US consumption of approximately 18 mbd in August 1990.[76] However, it helped calm the markets, by showing that a stock withdrawal would work in the case of a bigger emergency. This insurance was important psychologically, as government stocks had never been used before, and there was uncertainty as to whether their timely withdrawal was possible at all.

Apart from its apparently failed attempt to organize a coordinated stock release, the US role in leading the anti-Iraq effort is indisputable. Its indirect role for preventing a major oil crisis needs to be recognized. American influence could be noticed on all fronts—from Security Council debates through negotiations with Arab leaders to leading the military effort in January 1991. And it could not have been otherwise given US preponderance worldwide and its standing in the Middle East, in particular. Since the beginning of the 1970s, or the British Middle East withdrawal, the United States had gradually strengthened its diplomatic and military positions in the region. Besides the well-known "twin pillar" policy, successfully sustained until the Iranian shah's deposition, the United States also had military agreements with Omar and Bahrain. Egypt was the second biggest recipient of US aid, and got its American debt cancelled in return for its support for the coalition effort.[77] As a whole, a great variety of tools were used by the American government to induce cooperation. On 22 August 1990, after intense pressure on the Yemeni government, an Iraqi tanker stopped unloading its crude at the port of Aden.[78] Syria was lured into participating in the coalition effort by receiving billions of dollars of aid.[79] One of the most important moves of the American government, critical for both the outcome of the war and the oil crisis, was convincing Israel not to respond to the Iraqi Scud missiles attack on 18 January 1991. It is hardly imaginable how the conflict would have developed, had Israel gotten involved. It is very probable that Saudi Arabia and the other Arab exporters would have declared an oil embargo, producing a very severe shortage.

On the whole, while not able to coordinate the consumers' oil response, America's actions did prevent a much more severe shortage, and led to the successful resolution of the military crisis. The US leading role in the conflict needs to be examined in light of the singularity of the moment. At the time of the Iraqi invasion, the world community, and Europe in particular, was focused on several groundbreaking events. The Soviet bloc had just disintegrated, and the Soviet Union, itself, was about to collapse. Germany was in the process of unification, which was completed on October 3, 1990, and the first elections of the new state took place in December of that year. The Europeans, as a whole, were inward-looking at the moment, reflecting on their role in the new post–Cold War environment, and preparing for the Maastricht Treaty. The Yugoslav crisis was already brewing down south and the Soviets were preoccupied with the breakup and the transition to a market economy. This could explain the almost unconditional cooperation they offered, given their history of support to Iraq. The United States "was the only country that could respond" at the moment, being the winning power of the Cold War, and possessing immense diplomatic and military power.[80]

The IEA

In 1991, the IEA activated for a first time its emergency procedure. The Gulf War Contingency Plan, which made available 2.5 mbd, was put in place on 17 January 1991—the day the military intervention started—five and a half months after the start of the crisis, and three and a half months after the US government released oil from its reserves. The Plan put an emphasis on stockdraw, which was to account for four-fifths of the released oil. The rest was to be provided through demand restraint and fuel switching. Member-states had 15 days to arrange for oil to be released from their national reserves. The biggest contributor was supposed to be the United States, with 1 mbd.[81] The plan was well thought, however it came too late. In reality, member-states never put it into practice, as it was not necessary to intervene on the market after January.

Why did the Agency take so long to activate coordinated action, given that the 7% threshold was already reported in August? Meeting on August 31, 1990, the IEA Board took a decision not to activate emergency measures, since an immediate threat to supply did not exist. Stocks were high, and their sheer existence provided a psychological buffer. Production cuts were already offset. This was partially true, however, the price spike of October showed that oil sitting in caverns was not enough to calm the markets. In addition, the increased Saudi and UAE production took a long time to reach consumers. While waiting for this substitute, the lost Kuwaiti and Iraqi oil could have been replaced, at least partially, by reserves.

The IEA's January 17th decision received both criticism and praise. Not surprisingly, OPEC attacked the drawdown, blaming it for the price collapse on January 17, when oil dropped $10.56 down to $21.44. Some analysts argued it was unnecessary, given the oil market conditions at the time.[82] Indeed, by the time the Gulf War Contingency Plan came into effect, the market was already oversupplied. Increased production had made up for the lost oil, and tankers at sea were holding huge amounts of crude. The IEA reported that the amount of unsold oil “afloat and ashore” had actually increased at the end of January.[83] All this combined with reduced demand and energy conservation. Other market observers argued that the move came too late.[84] By acting earlier in the crisis, the IEA could have alleviated the October price spike, which was the most damaging episode of the war for oil markets and industrialized economies, as a whole.

The IEA’s actions in 1990 reflect the inadequacy of its emergency procedure. Although slightly reformed in the 1980s, it still remained focused on countering the physical shortage of oil and not the initial psychologically induced and panic-driven market behavior. This is surprising, given that most post–World War II shortages have been psychological rather than real. It is indicative of the unwillingness or inability of nation-states to forge concerted responses to international crises.

Unfortunately, there is no access to IEA archives from this period and historical records cannot be consulted at this point. Undoubtedly, there were disagreements among consumers on whether to release emergency stocks. The *New York Times*, citing the German economics minister, reported that West Germany and Japan opposed the use of oil reserves, unless there was a serious oil shortage.[85] It is hard to assess the German rationale in this case, but several important reasons need attention. Germany was more dependent on foreign oil than the UK and France, the former being an oil producer, and the latter—with a strong nuclear power industry. The Germans also had one of the biggest oil stocks in the world, which they might have been unwilling to share with less prepared countries. In addition, at the time of the crisis, the Soviet Union considerably reduced its oil deliveries to Eastern Germany, which put an additional burden to its Western neighbor. In any case, national considerations prevailed.

The role of the IEA January decision in lowering prices, although uncertain, needs to be recognized. It contributed to alleviating the panic, which might have ensued following the start of military action. All analysts predicted a considerable price increase if war were to break out. Fortunately, it did not materialize. However, it is difficult to assess the degree to which the IEA’s decision influenced the markets, given all the other factors playing at the time. Most critics attributed the price fall to a combination of factors, such as the success of the initial aerial attack, which left Saudi oil installations undamaged, the oversupply of the market at the moment, the Israeli decision not to retaliate to the Iraqi missiles, and the IEA decision.[86] As with most other events in international politics, it is difficult to determine the exact causal relationship. Undoubtedly, all of these factors played a role. Their degree of importance, however, is impossible to determine.

The First Gulf War: Conclusion

In sum, the 1990 oil shortage validates the power paradigm. Both the American inability to make others cooperate and its unilateral actions support its premises. The United States was not able to forge a cooperative response to the oil crisis. On the other hand, its unilateral stock drawdown helped ease market pressures. The United States had one of the biggest oil reserves in the world and showing its readiness to use it in case of an emergency was critical not only for the domestic, but for the international market at large. While America's efforts and leadership were instrumental for the successful outcome of the coalition military endeavor, its efforts within the IEA Board had a less favorable outcome. No other country could assume oil leadership at the moment. This resulted in divergences within the IEA and the belated activation of an emergency procedure in January 1991. At that time, the market had returned to normal and this action was unnecessary. This is probably why it was hardly implemented by member-states. The 1990 oil shortage did not develop into a full-scale emergency not because of the cooperative efforts of IEA member-states, but mostly because of the favorable market conditions at the time.

Hurricane Katrina

In 2005, for a first time, the IEA emergency scheme was activated and partially implemented by member-states. Ironically, this was done because of a shortage in the United States. Undoubtedly, at the time of the Agency's establishment in 1974, such an eventuality was not considered. The 2005 oil shortage is also interesting because it was caused by a natural disaster and not geopolitical factors, and also because it was the first "integrated energy shock," affecting simultaneously oil, natural gas and electricity.[87] Accordingly, while not treated as a separate case in this study, it deserves some attention.

Hurricane Katrina reached the United States Gulf Coast on 29 August 2005. It caused the disruption of 1.4 mbd of oil production and 1.8 mbd of refining capacity. The shortage of refining capacity was particularly threatening. The months before Katrina registered one of the lowest levels of spare refinery capacity, with refineries operating at the time at 99.5%.[88] Refinery outages were reported throughout July and August and prices increased with 22% between May and August.[89] Similarly to 1973, internal deficiencies preceded and magnified the shortage.

The global oil market, in general, was working at almost full capacity. Very strong demand growth in Asia, and particularly China, had deprived the market of the little spare capacity it had. World oil demand grew with the exceptional 3.4% in 2004, while supply lagged behind.[90] Naturally, prices went up and in June 2005, crude oil futures reached $60—a record high price in nominal terms.[91] The only element missing was a supply shortfall.

Then, Hurricane Katrina struck, halting the production of 3.2 mbd or 15% of US daily consumption. Gasoline prices went up and long lines were reported at some fueling stations.[92] The administration reacted quickly with President Bush declaring on 31 August that oil from the SPR would be lent to struggling refineries, and environmental standards would be temporarily relaxed. What really helped calming the market, however, was the IEA countries' decision on September 2 to release 2 mbd for the next 30 days. The United States was to provide half of this amount in crude, with the rest in the form of refined products from European governments and Japan. Some countries, notably the Netherlands, started selling reserve oil in the next several days. Almost all IEA member-states took part in the release and by early October 48 million barrels had already been added to the market. The decision brought immediate relief, as the NYMEX spot price fell one dollar the next day and continued dropping afterward.

On a first glance, the Katrina case clearly defies the power paradigm, which all other cases in this work support. The United States lacked spare capacity—the leverage it used pre-1970 to influence its allies. In all previous oil supply shortages international organizations reacted slowly. In the pre-1970 cases, their crisis response was largely US-dominated, while post-1970—their decisions remained only on paper. Why is it that in 2005, the International Energy Agency was able to quickly take a decision, and more than that—put it in practice?

The IEA's swift action is commendable and its role in taking pressure off the market needs to be recognized. There might have been some process of learning between 1991 and 2005, bringing a greater awareness of the need to react quickly. As there is no access to internal IEA documentation and discussions, this is difficult to assess. Upon closer examination of the Katrina events, however, several other important details transpire. First, the hurricane affected only American oil, and only for a short period of time. In fact, refineries started resuming operations the week after the hurricane struck.[93] European and Japanese oil supplies were, at no point, threatened. Hence, there was no need for Europeans to hoard and fill in inventories—a counterproductive behavior they engaged in during all post-1970 crises. Second, the nature of the shortage also helped, as it became clear early on that it would only affect American oil. A political supply disruption carries a much higher degree of uncertainty, whereby the duration of the crisis and its magnitude are hard to predict. It is this insecurity, which prompts market participants to compete for oil. It is highly unlikely that European governments would have agreed to release oil, if their own supplies were interrupted or threatened. Thus, the nature and magnitude of the Katrina crisis do not provide a good test for this book's argument, which was developed on the basis of observations of large oil supply shocks affecting most consumers.

Small Crises: Conclusion

The four smaller oil crises follow the same pattern as the big oil shortages of 1956, 1973 and 1978. The two pre-1970 cases saw higher levels of coordination, mostly due to strong American leadership. High spare capacity and early on planning, sending signals of reassurance across the Atlantic, were crucial in these cases. In 1951, the closure of the Iranian Abadan refinery led to the removal of more than 50% of European oil. Aviation fuel was particularly affected, and a major crisis was avoided because of the timely reaction of the US administration. Shortly after the refinery halted exports, the emergency committees were in place, production was increased, tankers rerouted and oil shipped to Europe. These factors were also important in 1967—the Six-Day War. There was one additional element that proved critical in this case: the American ability to persuade allies to coordinate efforts within the oil committee. This was a prime example of power in action.

The two post-1970 shortages developed differently. High levels of discord ensued in both cases, and prices increased significantly. During the onset of the Iran-Iraq war, the United States was not able to react and the International Energy Agency could not fill the vacuum. Prices increased by 150% and exacerbated the already severe recession. During the First Gulf War, America was the undisputed leader of the anti-Iraq coalition, but could not assuage market fears as it had done in the past. The shortage led to a 118% price increase. The market recovered quickly, however, due to the favorable conditions at the time, namely, high spare capacity and low demand.

These four cases come to validate the power paradigm. Cooperative behavior developed under strong leadership, and discord arose when leadership was absent. International institutions played a marginal role in all cases. The minor oil shortage caused by Hurricane Katrina in 2005 developed somewhat differently, simply because it was smaller in magnitude and affected only the United States. Hence, it does not provide a good test for the hypotheses advanced here. Fortunately, there has not been a major disruption in more than twenty years, and my argument cannot be tested in more recent conditions. There is a lot of speculation as to whether a major oil supply crisis could still happen, or if it did, whether its consequences would be as severe as in the 1970s. As I have discussed elsewhere in this work, two major camps stand out—the optimists and the pessimists. The former maintain that a big disruption is very unlikely, because of the current nature of the oil market, which is better suited to absorb shocks, political developments in the Middle East, the greatly reduced oil intensity of industrialized economies, and diversification. The pessimists respond that, on the contrary, oil production and spare capacity are increasingly concentrated, terrorist groups are more active than ever, and the oil market is becoming more competitive as new actors emerge.

I believe there is still a risk of an oil supply disruption. And if all consumers are affected, it is highly likely that they would engage in the same damaging behavior as before. Hurricane Katrina does not point to the opposite as it is not the type of case, with which this work is concerned.

Notes

1. Kapstein, *The Insecure Alliance*, 84.

2. Mostafa Elm, *Oil, Power, and Principle: Iran's Oil Nationalization and Its Aftermath*, 1st ed., Contemporary issues in the Middle East. (Syracuse, NY: Syracuse University Press, 1992), 36.

3. J. H. Carmical, "Foreign Oil Standby US Is Awaited," *The New York Times*, 22 October 1951.

4. *Foreign Relations of the United States, 1951*, 1985, Vol. IV. 1582.

5. Ibid. 1627.

6. "Minutes of the 9th Session of the OECD Oil Committee," (Paris: Organization for Economic Cooperation and Development, 1966).

7. Associated Press, "Joint Oil Compact Authorized by US," *New York Times*, 3 August 1951.

8. Kapstein, *The Insecure Alliance*, 83.

9. Ibid., 84.

10. J. H. Carmical, "Shortage Made Up in Loss of Iranian Oil," *New York Times*, 26 August 1951.

11. Ibid.

12. Associated Press, "Europe not Alarmed by Loss of Iranian Oil," *New York Times*, 24 August 1951.

13. L. A. Vincent and G. Matthys, "La Situation Economique," *Revue Economique* 2, no. 6 (1951).

14. "Minutes of the 60th Meeting of the OEEC Oil Committee," (Paris: Organization for Europan Economic Cooperation, 1951), 9.

15. Trevor N. Dupuy, *Elusive Victory: The Arab-Israeli Wars, 1947-1974* (New York: Harper and Row, 1978); Walter Laqueur, *The Road to War: The Origin and Aftermath of the Arab-Israeli Conflict, 1967/8* (Harmondsworth: Penguin, 1969); Nadav Safran, *From War to War: the Arab-Israeli Confrontation, 1948–1967: A Study of the Conflict from the Perspective of Coercion in the Context of Inter-Arab and Big Power Relations* (Indianapolis: Pegasus, 1969).

16. The peninsula was returned to Egypt in 1979 as part of the Camp David Peace Accords.

17. First reports of supply disruptions to Germany appeared in the press on June 21 – "Bisher Keine OECD-Beschlüsse," *Frankfurter Allgemeine Zeitung*, 21 June 1951.

18. United Press International, "Bedenken Gegen Ölembargo," *Frankfurter Allgemeine Zeitung*, 1 July 1967.

19. F. A. Z. Wilhelmshaven, "Erstes Rohöl aus Amerika," *Frankfurter Allgemeine Zeitung*, 25 July 1967.

20. Kapstein, *The Insecure Alliance*, 141.

21. High Level Group of the Special Committee for Oil, "Draft Preliminary Report on the 1967 Oil Emergency," (Paris: OECD, 1968), 2.

22. Ibid.

23. Kapstein, *The Insecure Alliance*, 144.

24. William M. Blair, "Declaration of an Oil Emergency Is Considered by U.S. and Companies," *New York Times*, 9 June 1967.

25. Oil, "Draft Preliminary Report," 33.

26. Ibid., 10.

27. "Minutes of the 3rd Meeting of the OECD Oil Committee," (Paris: Organization for Economic Cooperation and Development, 1962).

28. Oil, "Draft Preliminary Report," 5.

29. Ibid.

30. "Minutes of the 11th Session of the OECD Oil Committee," (Paris: Organization for Economic Cooperation and Development, 1967).

31. Oil, "Draft Preliminary Report," 13.

32. OECD Oil Committee, "The Outlook for Oil Supply and Demand," 2.

33. Brendan Jones, "Canal Closing Is Assessed," *New York Times*, 16 July 1967.

34. Kapstein, *The Insecure Alliance*, 197.

35. Yergin, *The Prize*, 711.

36. Ibid., 713.

37. Barry M. Rubin, *Paved with Good Intentions: The American Experience and Iran* (New York: Oxford University Press, 1980).

38. Bernard D. Nossiter, "U.S. and Allies Discuss Joint Fleet to Protect Vital Strait of Hormuz," *New York Times*, 25 September 1980.

39. Charles Percy, "To Meet an Oil Crisis," *New York Times*, 30 September 1980.

40. Paul Lewis, "Major Companies Plan to Cut Oil Purchases," *The New York Times*, 10 December 1980.

41. Kapstein, *The Insecure Alliance*.

42. Scott, *The History of the International Energy Agency*, 118.

43. John Tagliabue, "Spot Oil Price Climbing as War Pares Supplies," *New York Times*, 24 October 1980.

44. Kapstein, *The Insecure Alliance*, 198.

45. "Arabs to Hike Output To Offset Iran/Iraq Losses," *Oil & Gas Journal*, 13 October 1980.

46. "Saudis List Customers for Extra Crude," *Oil & Gas Journal*, 3 November 1980.

47. Kapstein, *The Insecure Alliance*, 198.

48. Michael Sterner, "The Iran-Iraq War," *Foreign Affairs* 63, no. 1 (1984).

49. U.S. Energy Information Administration, "World Crude Oil Production: OPEC Members," http://tonto.eia.doe.gov/merquery/mer_data.asp?table=T11.01a (accessed 15 March 2011).

50. Staff Report to the Committee on Foreign Relations, *War in the Persian Gulf: The U.S. Takes Sides* (Washington, D.C.: U.S. Senate, 1987).

51. Odell, *Oil and World Power*.

52. Daniel Badger and Robert Belgrave, *Oil Supply and Price: What Went Right in 1980?* (Paris: Atlantic Institute for International Affairs, 1982); Keohane, *After Hegemony*; Lieber, *The Oil Decade*.

53. Yergin, *The Prize*, 713.

54. Ibid., 706.

55. Tim McGirk, "How Far Will Iraq Go," *Christian Science Monitor*, 25 September 1980; "War Curbs Oil Exports by Iran, Iraq," *Oil & Gas Journal*, 29 September 1980; "War in the Oil Fields of Iran and Iraq," *Newsweek*, 6 October 1980.

56. "The Community and Events In The Middle East," http://aei.pitt.edu/5582/01/002280_1.pdf (accessed 15 January 2011); "Transcript of Carter's Statement on Iran-Iraq War," *New York Times*, 25 September 1980.

57. See, for example, Dilip Hiro, *Desert Shield to Desert Storm: The Second Gulf War* (New York: Authors Choice Press, 2003); Majid Khadduri and Edmund Ghareeb, *War in the Gulf, 1990–91: The Iraq-Kuwait Conflict and Its Implications* (New York: Oxford University Press, 1997); Ken Matthews, *The Gulf Conflict and International Relations* (London; New York: Routledge, 1993); Steve Yetiv, *The Persian Gulf Crisis* (Westport, CT: Greenwood Press, 1997).

58. International Energy Agency, *Monthly Oil Market Report: April* (Paris: IEA, 1991), 25.

59. Ibid.

60. International Energy Agency, *Monthly Oil Market Report: October* (Paris: IEA, 1990).

61. Hiro, *Desert Shield to Desert Storm.*

62. International Energy Agency, *Monthly Oil Market Report: October.*

63. Lowell S. Feld, "Oil Markets in Crisis: Major Oil Supply Disruptions Since 1973," in *The New Global Oil Market: Understanding Energy Issues in the World Economy*, ed. Siamack Shojai (Westport, CT: Greenwood Publishing Group, 1995), 113.

64. Robert D. Hershey, "War Effect on Energy Minimized," *New York Times*, 28 December 1990.

65. Matthews, *The Gulf Conflict and International Relations*, 204.

66. Francois Heisbourg, "France and the Gulf Crisis," in *Western Europe and the Gulf*, ed. Nicole Gnesotto and John Roper (Paris: Institute for Security Studies of WEU, 1992), 21.

67. International Energy Agency, *Monthly Oil Market Report: January* (Paris: IEA, 1991).

68. Hiro, *Desert Shield to Desert Storm.*

69. "Third Time Lucky?," *The Economist*, 6 October 1990.

70. International Energy Agency, *Monthly Oil Market Report: October*, 1; Robert J. Lieber, "Oil and Power after the Gulf War," *International Security* 17, no. 1 (1992): 172.

71. Hiro, *Desert Shield to Desert Storm*, 106.

72. Ibid., 156.

73. Yergin, *The Prize*, 770.

74. U.S. Congress. House, *Energy Impact of the Persian Gulf Crisis: Hearing Before the Committee on Energy and Commerce*, 101 Cong., 2 sess., 1990.

75. N. Nash, "Pressure Is Mounting to Tap U.S. Oil Stockpile," *New York Times*, 18 September 1990.

76. U.S. Energy Information Administration, "U.S. Product Supplied of Crude Oil and Petroleum Products."

77. Tibi, *Conflict and War in the Middle East*, 163.

78. Hiro, *Desert Shield to Desert Storm.*

79. Tibi, *Conflict and War in the Middle East*, 153.

80. Matthews, *The Gulf Conflict and International Relations*, 98.

81. International Energy Agency, *Monthly Oil Market Report: April.*

82. Simon Beavis, Ben Laurance, and Deborah Wise, "Finance and Economics: War in the Gulf—International Energy Agency Activates Contingency Plans by Releasing Extra 2.5m Barrels Day on to the Market." *The Guardian*, 18 January 1991.

83. International Energy Agency, *Monthly Oil Market Report: February* (Paris: IEA, 1991).

84. Lieber, "Oil and Power after the Gulf War," 174.

85. William Drozdiak, "International Agency Calls Oil Supplies Adequate," *Washington Post*, 29 September 1990; Editorial, "IEA Snubs Bush on Strategic Oil," *Oil & Gas Journal*, 8 October 1990; Youssef Ibrahim, "Bush's Move on Stockpile Brings Some Dissent Overseas," *New York Times*, 28 September 1990.

86. Martin Barrow, "Price of Petrol Under Pressure as Oil Plunges," *The Times*, 19 January 1991; Beavis, Laurance, and Wise, "Finance and Economics: War in the Gulf—International Energy Agency Activates Contingency Plans by Releasing Extra 2.5m Barrels Day on to the Market"; "Leading Article: Volatile Oil," *Lloyd's List*, 18 January 1991; Mark Potts and Thomas W. Lippman, "Oil Prices Tumble on World Markets,"

Washington Post, 17 January 1991; Matthew L. Ward, "Price of Oil Dips Below $20 a Barrel," *New York Times*, 19 January 1991.

87. Daniel Yergin, "Ensuring Energy Security," *Foreign Affairs* 85, no. 2 (2006): 70.

88. U.S. Energy Information Administration, "Refinery Utilization and Capacity," http://tonto.eia.doe.gov/dnav/pet/pet_pnp_unc_dcu_nus_m.htm (accessed 10 March 2011).

89. U.S. Energy Information Administration, "This Week in Petroleum," http://tonto.eia.doe.gov/oog/info/twip/twiparch/twiparch.html#2004 (accessed 10 March 2011).

90. U.S. Energy Information Administration, "International Energy Outlook," http://www.eia.doe.gov/oiaf/archive/ieo05/index.html (accessed 10 March 2011).

91. Ibid.

92. Vikas Bajaj and Anthony Ramirez, "Spot Shortages of Gas Reported Around Country," *New York Times*, 5 September 2005.

93. Sam Fletcher, "Production, Refining Resumes Slowly Following Katrina," *Oil & Gas Journal*, 12 September 2005.

Chapter 7
Oil Crises and Inter-state Cooperation: Is It Possible?

This volume set itself the uneasy goal of analyzing consumer behavior during oil shortages. The study is important not only because of the continuing relevance of oil in our lives, but also because it addressed one of the most enduring debates in international relations theory—the one between neorealism and neoliberalism. Both perspectives have elaborated extensively on the issue of international cooperation—one of the most salient international issues, indeed.

In this book, two propositions, advanced by these two contending schools, were contrasted:

1) Nation-states are more likely to cooperate under the leadership of a dominant power. As neorealists argue, the latter has the resources both to coerce others to engage in cooperation, and to serve as a supplier of last resort.
2) International institutions can and do facilitate cooperation. As institutionalists predict, IOs provide a forum for debate and the exchange of information, decrease transaction costs and increase opportunities for issue-linkage. All these factors make cooperation more likely.

So Who is The Winner?

The post–World War II period provided a rich testing ground for the above hypotheses. Within a time span of fifty years, the world witnessed seven major oil supply disruptions—the Abadan crisis (1951), the Suez crisis (1956), the Six-Day War and embargo (1967), the Yom Kippur War and embargo (1973), the Iranian Revolution (1978–79), the Iran-Iraq War (1980) and the First Gulf War (1990–91). All these cases were taken into consideration in the present study. The crises of Suez, the Yom Kippur War and the Iranian Revolution were examined in detail in chapters devoted to each of these events respectively. The other four cases and Hurricane Katrina were surveyed briefly.

The evidence seems to be supporting the neorealist perspective. European states successfully coordinated their emergency responses to the crisis in the cases where the United States was able to exert power and to serve as a supplier of last resort. International institutions helped only marginally and only in those cases where the United States provided leadership.

The cases examined were carefully chosen on the basis of the amount of oil withdrawn from the market, and their economic impact. The selection also allowed me to test four separate international organizations, which were active in different periods. The OEEC oil committee was the center of policy coordination in the 1950s. It was formed with the goal of assisting with the allocation of Marshal Plan money, and its main responsibility was to coordinate European refinery capacity expansion. The committee did not have emergency coordination powers, however it got involved in crisis policy coordination during the Suez crisis. The OEEC was succeeded by the Organization for Economic Cooperation and Development in 1961. The oil committee's functions remained largely unchanged with one important exception: an oil emergency procedure was introduced in 1962. Based on the 1956 experience, when oil reapportionment was in the center of the crisis effort, the scheme focused on oil allocations. Ninety percent of available supplies were to be distributed among countries on the basis of consumption levels, the remaining 10%—allocated to the neediest, or left for particular circumstances. This scheme remained in place until the establishment of the International Energy Agency in 1974, which took over emergency planning. It was never put to the test, however, as in 1967, oil reapportionment did not take place, and in 1973, there was no coordination at all. The IEA was the first full-fledged international organization devoted to energy issues. Its emergency program included three components: emergency stocks, demand restraint measures and as a last resort, allocations. After the dismal record of the IEA during the Iranian revolution, oil allocations gradually lost their significance, and emphasis was placed on coordinated stockdraw. This system was put in practice for the first time, although not very successfully, during the First Gulf War in 1991.

In several cases, crisis behavior within the European Community was also examined. The EC adopted its first emergency measures in 1968, in the aftermath of the Six-Day War embargo. Therefore, coordination within the community was considered in the cases following that crisis. The EEC provides an important test as it presents the highest level of international institutionalization. It is a peculiar body resembling a typical international organization in some ways, but also one in which nation-states have pooled resources to a much higher level than is usually the case. The inclusion of the EC in the study increases its validity, as the organization gives a very high value on one of the independent variables (IOs). Hence, if the study variable—cooperation—also gives high values, when measured against the EC, then the tested theory holds true. The theory in this case is neoliberal institutionalism. Unfortunately, this was not the case.

Two important points need to be made regarding international institutions. First, in all cases, pre- and post-1970, international organizations reacted slowly. This is not surprising, given their intergovernmental nature, where statements have to be consulted with capitals, decisions have to be taken with unanimity, and the organization machinery, in general, is more rigid than a governmental one. However, after 1967, these institutions had emergency schemes in place. An emergency requires fast reaction capabilities, which these bodies did not have in any of these cases. In contrast, the swiftness of the American government crisis response was impressive in all pre-1970 cases.

Second, after 1970, in the cases when crisis-related decisions were taken, these were not implemented by member-states. The International Energy Agency's attempts at coordinating demand and spot purchasing restraint in 1978 and 1980 remained largely unheeded. Member-states followed policies in tune with their immediate needs, which basically consisted of securing whatever oil supplies were available. In 1991, the Agency adopted the Gulf Contingency Plan, aimed at increasing oil availability on the market and decreasing demand. While often evoked as one of the successes of the IEA, the plan did not materialize as it was adopted too late, after the market had returned to normal.

The other critical variable—power—was defined as United States resources in the oil issue-area. During the evaluation of the evidence it became clear that among these resources the most important were American spare production capacity, early on emergency planning, and the use of persuasion and coercion over European allies. The first two components were especially important, as they provided a psychological buffer, serving as a sort of an oil market hedge. Early on crisis preparations sent positive signals to allies, who were ensured they would be provided assistance, and therefore did not have to scramble for or hoard supplies.

In several of the cases examined, the United States did use all three factors. In these cases, the highest level of crisis policy coordination was achieved. These were the Abadan, Suez and the Six-Day War crises. Not surprisingly, these cases coincided with the postwar period of American supremacy in oil, as measured by its spare capacity, production-to-consumption ratio, and American oil companies' control over the market. The American power of persuasion was brilliantly used in 1956—to coerce allies to work together in the reallocation of supplies, and in 1967—to convince the OECD oil committee to declare an emergency. In all pre-1970 cases, the American administration organized an emergency response within the Foreign Petroleum Supply Committee within less than a month from the start of the crisis and granted antitrust exemption to its oil companies, which then were able to coordinate European oil supply. The companies increased output in oil–producing states (mostly Texas), put mothballed tankers into operation, and rerouted other tankers toward European ports. All this was achieved with close coordination with the American government, European oil companies (Shell and BP) and, in 1956 and 1967, European governments within the OEEC and OECD oil committees, respectively.

The period of American dominance in oil ended approximately around 1970. That year, the United States oil production reached its peak, and started declining afterward, while consumption kept growing. Imports increased dramatically, with the steepest increase registered between 1967 and 1973—from 19 to 35% of consumption. Spare production capacity reached zero in 1971 and has remained at that level or insignificantly low since then. As this critical factor was missing, there were no early crisis preparations on the American side after 1970. The decline in American power in the oil issue-area coincided with the highest levels of crisis discord among European consumers. The lack of cooperation magnified the crises' economic consequences, as consumers rushed to search for supplies, hoarded oil and signed very expensive bilateral oil contracts. The 1973 oil crisis was particularly severe and within three months prices rose 400%—the highest peacetime increase ever in such a short period of time. In 1978 and 1980, the price increase was of 150%. In comparison, during the pre-1970 crises, prices increased between 30 and 50%.

The international institutions also did better during the pre-1970 period. In 1951, the OEEC oil committee limited itself to information-sharing, as the timely and efficient American reaction obviated the need for action on its part. The 1956 crisis saw the highest IO involvement. The oil committee participated in oil supplies reallocation in coordination with its industry board—the OEEC Petroleum Emergency Group. This was also the only time oil reapportionment took place. The committee's assistance was very effective, however, it was only activated in January, after the American companies had mobilized resources and had almost filled the European oil shortage. In 1967, the committee's initial refusal to act was turned around, after the United States threatened not to ship the needed oil, unless committee members declared a state of emergency. This declaration paved the way for the establishment of the industry advisory board, which coordinated the emergency supplies in conjunction with the American industry body.

The post-1970 story is more versatile. In 1973, the OECD and EEC members, eager not to antagonize Arab producers, did not react to the crisis at all. The severe economic damage brought by the oil shortage prompted the establishment of the International Energy Agency. Created with high expectations, its performance in 1978 was disappointing. The IEA Board met for the first time in March 1979—almost three months after the complete halt of Iranian production. Member-states agreed to reduce oil consumption by 5% of projected 1979 levels. The method of reduction was to be decided by them without any accountability and follow-up. The decision remained only on paper, and some members even registered increased consumption. The EEC shared a similar fate, agreeing in March to reduce demand, and taking no action afterward.

In 1980, the IEA Board reacted slightly faster, advising member-states as early as October 1980 to refrain from abnormal purchases on the spot market, and to use their stocks. This was the Agency's only action during the crisis and it was not implemented by member-states. Ten years later, during the First Gulf War, the IEA, for the first time put in place a coordinated plan of action. It involved a coordinated stockdraw, demand restraint and fuel-switching. The plan was well thought, however, it came as late as January or five months after shortages were reported, and accordingly, was not implemented, as there was no need to act at this point. Table 10 below provides a summary of all the cases.

Table 10. Summary of the cases

crisis	US reaction (time after the start of the crisis)	US production-to-consumption ratio (percentage)	US antitrust exemption	Emergency petroleum committee in the United States	Foreign involvement of the US committee	IO involvement
1951	within two weeks	95	X	X	X	none
1956	3 months prior to crisis	90	X	X	X	allocations
1967	2 days	81	X	X	X	consultations
1973	none	63		X		none
1978	7 months	55				none
1980	none	59				none
1991	1 month	54	X			emergency plan

The empirical evidence, collected and evaluated on the basis of these cases, gives support to the hegemonic stability thesis. In the pre-1970 cases, Europeans were more successful at crisis coordination, mostly due to American leadership. This leadership materialized in the form of spare production, refinery and tanker capacity, oil companies' influence over the oil market, and the ability of the administration to organize quickly and to exert influence in the respective international body. Around 1970, American oil production reached its peak, and wells started pumping oil at full capacity. At the same time, oil companies lost some of their leverage after a wave of nationalizations. These developments had important implications for cooperation. The psychological cushion provided by American reserves and capacity was gone. This considerably increased the uncertainty and consumers opted for self-help policies. As expected, these policies magnified the shortage's nefarious effects on the economy.

The latest case of an oil supply shortfall—Hurricane Katrina in 2005—was also briefly surveyed. It is an interesting case, as it provides the only instance of post-1970 policy coordination. Several days after more than 2 mbd of crude and products were removed from the market, the IEA adopted a plan of action. Member-states were to release 2 mbd for the next thirty days. Most European governments and Japan participated in the stockdraw, which helped bring the price down.

While seemingly in contradiction with the power hypothesis, the Katrina case needs to be seen in a different light. In the first place, the hurricane affected only American oil and only for a short period of time. European supplies were never threatened, and withdrawing oil to calm the markets was never controversial. In the second place, the nature of the crisis was different than any of the other cases. Uncertainty is much higher when the shortage is caused by political disturbances, which duration and magnitude is hard to predict. This study looked at big supply shortages, affecting all or most consumers. It is highly doubtful that coordinated action would have taken place, had Katrina been one of those cases.

Theoretical Implications

This study falls within one of the great debates in international relations theory—between the two mainstream theoretical approaches—neorealism and neoliberalism. The ability of international actors to cooperate for the achievement of common goals and their rather more prominent inclination toward discord have been thoroughly studied by IR scholars. The importance of this field of research is underscored by the paradoxical nature of the cooperation problem—it is well known that cooperation leads to win-win solutions, and yet, it is so evasive. In a world populated by rational actors, this should not be the case. And indeed cooperative behavior does occur in some cases. The question examined in this volume is under what conditions cooperation emerges. Two competitive explanations of the cooperation phenomenon have been advanced by the above theories—the hegemonic stability thesis and institutionalism. Hegemonic stability theorists maintain that cooperative behavior can only occur under the leadership of a dominant power. In addition, institutions are only effective as long as they are supported by this power. The leader has the ability to coerce others, and to provide resources and serve as a supplier of last resort. Institutionalists counter this argument by saying that international organizations can foster cooperation even in the absence of a leader, by providing a forum for debate and exchange of information, reducing transaction costs and increasing opportunities for issue-linkage. The area of oil crisis cooperation provides important evidence, which lends support to the hegemonic stability thesis.

Applying these theoretical perspectives to the issue-area of oil is important, as oil crosses the barrier between security and economic issues. There exists a security—political economy divide in the study of international relations, and the area of cooperation, in particular. Neorealists have often argued that international organizations can only be effective in the coordination of "low politics" issues. Institutionalists have countered that "low" and "high politics" issues have the same cooperation impediments—lack of information and cheating, which could be overcome with the help of institutional mechanisms.

Issues of oil supply and demand sit at the intersection of the security and economy fields. Oil is a commodity like any other, traded on international markets. Its prices are determined by the laws of supply and demand. At the same time, it is a commodity, which is vital for states' survival, which makes its regular supply at affordable prices one of states' prime goals. This allowed testing the theories in a sort of a combined setting, including features of both the security and political economy environment.

By comparing these two dominant IR perspectives and eliminating one of them as a possible explanation for the incidence and conditions of cooperation, I performed a "crucial test."[1] According to Stinchcombe, eliminating a theory's most probable contender increases the validity of the theory. Similarly, Grieco argues that "the most powerful way to test a theory is to determine if the propositions derived from it hold in circumstances…in which comparable but divergent propositions from competing theories very much ought to be validated."[2]

The propositions made by these theories were examined in seven cases of oil supply disruptions by using the method of structured, focused comparison. The use of such a controlled comparison method allows for the systematic study of historical cases by taking them out of their specific context, and comparing carefully chosen features of each. In this way, possible biases or third variable influences are avoided as much as possible. This increases the reliability of the study and its findings.

While the power paradigm is the clear winner of this case study comparison, caution is always needed when using hegemonic stability theory as an explanatory tool. The theory takes only tangible factors into account, but intangibles are also very important, for example, the US–Saudi Arabia relations or US position in the Middle East, in general. Despite the fact that the United States does not have a leadership position in the oil issue-area anymore, it might still be able to exert influence through its partnership with Saudi Arabia. The role of intangible indicators of power might be a good venue for future research.

Policy Implications

This study is the first attempt to compare consumers' crisis cooperation across seven post–World War II oil shortages. It is an important step toward our understanding of the dynamics of oil supply shortages and inter-state cooperation. Oil is a critical commodity and will remain so for the foreseeable future, accounting for 30% of the energy mix in 2030.[3] Natural gas use is also expected to rise. While renewable energies will be gradually making their way, fossil fuels' use will also increase. By 2030, these will be taking 80% of our energy use.[4]

In the meantime, oil will be increasingly concentrated in a handful of unreliable regimes. Saudi Arabia holds the biggest oil reserves, estimated currently at 266.8 billion barrels.[5] In addition, it has most of the world's spare capacity—3.76 mbd in 2010.[6] Russia, while not so abundant in oil, is currently the biggest oil provider to EU countries, supplying 33% of their needs. This dependency is projected to increase. Russia has been a very divisive factor in EU politics in the last decade. The Russian government has preferred to approach European countries bilaterally, which has led to a number of under-the-table deals and a general lack of transparency. This contributes to an atmosphere of mistrust and competition among European countries in the field of energy supplies.

In addition, the world of oil is becoming increasingly competitive, as more consumers enter the market. China, which was self-sufficient in oil as recently as 1993, is now the second oil consumer after the United States. This dramatic rise in Asian oil demand brought a price increase of more than 100% between 2004 and 2006. The price march up was only stopped by world recession in 2008, but is expected to resume after the economy recovers. China's energy requirements are projected to increase by 150% by 2020.[7] In order to satisfy its growing oil hunger, China has acquired interests in production in all major exporting countries, including Saudi Arabia, Iran, Venezuela, Russia and Canada. So far, the Chinese have shown little interest in cooperating in the oil issue-area. On the contrary, their disposition has been one of competitiveness. In addition, China is not a member of the International Energy Agency, although some form of a dialogue between the two has been established.

The findings of this study, while especially relevant for oil, which currently is the most important traded commodity, could be extended to other natural resources. Natural gas is the prime candidate, even though the natural gas market is organized differently from the oil market. There are two important differences between the two markets. First, it is a regional market, where most gas is transported through pipelines, and thus cannot cover very large distances. Second, it is organized on the basis of long-term contracts, as opposed to the spot "nature" of oil. However, it is exactly these features that make natural gas supplies vulnerable to disruptions. Gas is much harder to transport than liquid oil, as 95% of it runs through pipelines, and its destination is known before it is even produced.[8] Thus, an interruption of gas supplies to a particular country could not be easily countered, as in the case of oil. Most of the oil shortages, discussed here, were quickly covered by increased output in other producing countries. This would be difficult to accomplish with natural gas, with the exception of the relatively small market of liquefied natural gas (LNG). In addition, due to the rapid expansion of the LNG network, the natural gas market is slowly changing and starting to acquire some of features of the oil market.

Europeans are very much aware of their vulnerability in this area as they have suffered a number of natural gas disruptions in the not so distant past. This vulnerability will only grow in the future, as natural gas reserves will be increasingly concentrated in Russia. Currently, Europe receives natural gas from a number of locations including Norway, the Netherlands and North Africa. Norwegian and Dutch production has already reached its peak. Russia, on the contrary, holds 25% of the world's reserves, standing at 1,577 trillion cubic feet. In comparison, Iran—which has the second biggest gas reserves—holds about 980 trillion cubic feet. European dependency on Russian gas varies, from country to country, with some Eastern European states importing 100% of their needs from Russia. Germany is dependent on Russia for 40% of its gas, and this number is expected to rise to 60% by 2020. This increasing dependence poses new risks, as the Russian administration has repeatedly used supplies as a tool of coercion. A Swedish energy security study has identified fifty-five politically motivated supply cuts since the breakup of the Soviet Union.[9] One of the most severe disruptions—in January 2009—caused electricity and heating cuts in Eastern Europe and was felt as far as France and Italy. Russian natural gas supplies will keep being the Achilles' heel of Europe for the foreseeable future.

The findings of this study are especially important for international cooperation, and international organizations, in particular. The number of international institutions has significantly increased in the past couple of decades. Almost every issue-area of international life is organized on an intergovernmental level. As a consequence, there are high expectations that these institutions will deliver on their most important goal—providing cooperation. This study, however, shows that these expectations might be exaggerated. At least in the case of issues critical for states' survival. We either have to improve the existing institutional mechanisms or embrace the power reality of international politics, and adjust our behavior to its opportunities and constraints.

What Does the Future Hold?

It is difficult to predict whether a major oil supply disruption will happen, let alone when. Oil analysts have been notorious for their inaccurate forecasting, even when it comes to more calculable variables like prices, and supply and demand patterns. For example, a number of positive projections were made in the late 1960s—several years before the catastrophic 1973 shortage.[10] Similarly, after the Second Oil Shock, analysts predicted another big crisis in the mid-1980s or recurrent oil shocks in the future.[11] While some market watchers and scholars have lately asserted that oil shortages are a phenomenon of the past, these reports should be taken with caution, given our past failures to accurately assess the oil market. Based on the results of this study, two possible scenarios come to mind. The first one concerns a real supply shortage occurring in times of a tight demand-supply balance and removing substantial quantities of oil from the market. While the probability of this happening is relatively small, if it materializes, the current emergency system is unlikely to respond adequately. If all major consuming hubs are affected, it is very unlikely that any form of cooperative behavior will ensue. The crises of 1973 and 1978, and to a certain degree 1990, are good examples. If the physical flow of oil is interrupted, it is in states' primary interest to ensure whatever supplies are available.

The second scenario concerns either a smaller interruption of supplies affecting only one region or country, or a threat of such an interruption. A supply shortage in one country could be offset by increased production in other places or by stockdraws. This was the experience during Hurricane Katrina or to a lesser degree, the Venezuelan strikes of 2003. In this case, emergency measures in less affected countries could calm market fears and reestablish some balance. Cooperation is also more likely to emerge if only one or several consumers are affected.

In the case of the second scenario, other factors could also intervene and alleviate the shortage. In the first place, market forces need to be recognized as a sort of cushion, as well. The oil market, in its current form, developed in the early 1980s, when NYMEX introduced oil futures. Gradually, these became the most traded futures. The oil market became a separate entity, with its own life and rhythm. It undermined OPEC's influence on prices and, to a degree, increased predictability. These developments coincided with price deregulations in most consumer countries. Reagan's deregulation of oil prices in 1981 was followed shortly by similar moves in France and other European countries. Free market policies were in vogue.

The market could be relied on to absorb shocks and sustain a demand-supply balance in cases of smaller shortages. If prices go up, demand will shrink accordingly, bringing prices down. This was the case in the mid-1980s, when prices dropped to one of their lowest points ever, as a result of the price spikes of the 1970s. Producers had to adjust, as well, and Saudi Arabia reduced production from 10 mbd in 1979 to 2.3 mbd in 1985. More recently, demand fell as a result of the mid-2000s price increase. These natural adjustment processes could be relied on in the long run to balance the market and provide a degree of security.

However, oil price elasticity is rather low and market adjustments occur slowly. Major supply disruptions, that is, our first scenario, would cause considerable damage and take their social toll before equilibrium is restored. It is in these cases that cooperation will be more critical, and it is for this scenario that the international emergency system must be adapted.

Currently, our emergency preparedness is based mostly on a stock requirement. IEA member-states are required to hold stocks equal to 90 days of net imports. A similar requirement exists within the EU. Stocks are believed to provide a buffer against both domestic and foreign shortages—psychologically, but also by providing oil to an undersupplied market. Their significance has increased with time and they are currently considered the backbone of the IEA oil emergency system. Indeed, the amount of stocks has grown through the years. IEA countries' stocks were 3 billion barrels in the mid-1980s and grew up to 4 billion in less than twenty years. In addition, the proportion of government-held stocks, as opposed to industry, has increased. The United States holds the biggest and most developed oil reserve, amounting to 726 mbd or 38 days of consumption. In total, OECD countries hold about 4.1 billion barrels of government or agency/industry stocks, equal to 47 days of current demand. Even if only the 1.5 billion barrels of public stocks are taken into account, a coordinated action could supply significant volumes of oil over long periods. At a drawdown rate of 2 mbd, public stocks would cover 24 months.

Stocks would indeed provide an efficient buffer against supply shocks, if only they were used. The only time when IEA member-states released stocks in a coordinated way was during Hurricane Katrina. As already discussed, this was not a hard decision, as most consumers were not affected by the shortage. In 1991, during the First Gulf War, the IEA adopted the Gulf Contingency Plan, consisting of coordinated stockdraw, demand restraint and fuel-switching. The plan was not implemented, as by the time it was put on the table, the market was oversupplied and action was not necessary. In most other cases, consumers filled in inventories during shortages and used them during times of normalcy—a counterproductive behavior, which defies the stocks' purpose.

There is still no guarantee that in case of a major supply shortfall, stocks would be used. Although some process of learning might have taken place, past shortages indicate that their use cannot be relied on. This is especially relevant for big shortages, affecting most consumer countries. Let us recall that psychological factors are usually more important than the physical shortage. High uncertainty as to the duration of the crisis and its magnitude would make any rational leader cautious to use oil inventories. When their survival is threatened, states tend to fend for themselves, opting for self-help policies, as opposed to coordinated action.

In addition, while the West is relatively well prepared for emergency, the rest of the world is lagging behind. And it is actually the rest that will be growing in demand in the future, while OECD oil use is projected to decrease. China—the biggest contributor to demand growth—is a good case in point. China started developing strategic reserves in 2004, and currently holds about 100 million barrels or 12 days of consumption. The government has a plan of reaching the 60-day threshold by 2020. Whether it will participate in any emergency coordination, however, is questionable. Western governments have a tradition of working together and an emergency system, which although not perfect, has been tested and might be effective in countering smaller disruptions. In contrast, the Chinese could be described as rather difficult partners. So far, they have shown little tendency to cooperate on energy matters. And China is not the only critical actor on the oil market. Other growing economies, like India and Brazil, will contribute greatly to oil demand growth in the future. Cooperation with them on emergency stockdraw is everything but not certain. These countries are not part of the IEA, and coordination with them has been very slow to develop.

In sum, oil stocks provide a good security cushion, if used in a coordinated effort. Whether such coordination will occur in the case of a major supply disruption, is not certain. Past evidence suggests that when faced with an emergency, states prefer unilateral action. In addition, emerging oil actors do not have sufficient inventories and are not integrated into the international emergency structure. This makes cooperation with them even more doubtful.

The third factor, which is sometimes considered critical in providing energy security, is world spare capacity. Spare capacity did, indeed, play a role in some of the cases. For example, the 1980 shortage was much shorter and less severe than the one a year earlier, because of the high spare capacity in Saudi Arabia and a number of other producers, which were able to offset the shortfall very quickly. It did not, however, prevent hoarding behavior, and did not foster cooperation.

American spare capacity was one of the variables examined in this study. Clearly, it was shown to be one of the most important factors in all pre-1970 crises. Increased production in Texas, combined with good coordination between government and companies, and reorganization of oil shipping were very critical. Currently, most spare production capacity is concentrated in Saudi Arabia. The number is changing depending on demand growth, but it has been around 3.04 mbd for the year 2009.[12] This represents 3.8% of world consumption. There is no agreed security threshold of spare capacity. One thing is sure, however—that this capacity can very quickly disappear. In 2004–2005, following several years of exceptional growth in consumption, spare capacity reached 1%. This could happen again, given that oil demand in China and other emerging economies has already recovered its upward trend. In general, it is expected that demand growth will continue outstripping new production. In addition, capacity will be increasingly concentrated in Saudi Arabia. Although currently the Saudis produce only 14% of global oil, their share is expected to increase, as they hold a dominant proportion of world reserves.[13] High concentration, in the hands of a not so reliable regime, only increases consumers' vulnerability.

Besides pure numbers, perceptions are important, as well. United States oil power was critical for allies' security. It was the certainty of American assistance that calmed European fears. As long as America was able and willing to help, Europeans were certain to get the needed oil. There is no such guarantee with Saudi Arabia or any other Middle Eastern producer, for that matter, and production in Western countries is in decline. In addition, the power exerted by the US in its time of primacy, consisted of more than spare capacity, but also control over the market, and a dominant share of world production, among others. No other actor has ever had that level of influence on the market, in addition to being part of the Western block. This has important implications for Western energy security. In sum, world spare capacity might provide some level of supply shock buffer, however its existence, in general, is not guaranteed and it will be increasingly concentrated.

One important variable that might, actually be more critical than spare capacity is US military presence in the Persian Gulf, and American relations with Gulf countries, in general. United States power in oil has declined, however American global reach, in terms of its economic and military determinants, has remained and even increased. The role of American global leadership is a matter for a separate study. However, a few important points deserve attention here.

There is a general agreement that America has vital interests in the Middle East, although, presently, it receives only a fraction of its oil from the region. European countries and Japan are much more dependent on Gulf imports, with the latter receiving close to 80% of its oil from there. American presence in the Gulf dates back to the early 1970s, when the UK officially withdrew from the region. Over the years, the American Gulf presence gradually increased. These capabilities now include both military personnel and equipment, and a number of military bases, such as the Fifth Fleet base in Bahrain. In addition, the United States has greatly improved its relations with Gulf States in the past couple of decades.[14] All this played in American favor during the First Gulf War, where the United States emerged as the sole global power capable of projecting capabilities in the region.

Without a doubt, American presence in the Gulf is important for allies' energy security. There are a number of factors, however, which might escape the US shield. First among them is the threat of a terrorist attack. Small-scale terrorist attacks on oil facilities have already happened several times. For example, in May 2004, an Islamic extremist opened fire in the petrochemical complex in Yanbu, while shortly after, twenty-two foreign oil workers were killed in the city of Khobar.[15] It is a matter of time for a bigger terrorist attack to happen. Second, Europe currently receives 33% of its oil from Russia—a number that is expected to increase. In the last decade, Russia has several times surpassed Saudi Arabia as the first world oil producer. The only guarantee against a Russian oil supply disruption is that Moscow needs to sell its oil as much as Europeans need to buy it. This interdependence is important. However, Russia is known for using energy as a foreign policy tool, and there is no guarantee against that. Europe also imports some oil from Central Asian regimes, close to Russia.

Third, political and economic disturbances happen in other producing regions, as well. Let us remember the Venezuelan strikes in 2002 and 2003, which removed about 3 mbd from the market. This affected mostly American imports, as Europe consumes very little Venezuelan oil. However, it contributed to the price rise of that period. Venezuelan production has not completely recovered yet, and currently stands at some 500,000 b/d below pre-strike levels. Another important American supplier—Nigeria—is periodically struck by violence and production closures. The continent of Africa, as a whole, is believed to contain large amounts of undiscovered oil, most of it, in very volatile or even dangerous locations.

All these factors might have important implications for oil politics and for the daily functioning of the oil market. Some of them could also be important for cooperation. If market conditions are favorable and spare capacity is high, there is a higher chance that consumers will coordinate their emergency responses. Even under these conditions, however, if the shortage is substantial and affects all consuming countries, it is highly questionable that they would cooperate. Previous major supply disruptions support this conclusion.

Concluding Thoughts

The issues discussed in this volume will continue to be important in international politics. As we keep and even deepen, as expected, our reliance on fossil fuels, their regular and affordable supply will be increasingly critical. At the same time, inter-state cooperation will remain a divisive and puzzling topic. There is widespread agreement that cooperation can bring solutions to distributive dilemmas, and to crisis situations, in general. Under what conditions do actors cooperate, however, will be the most difficult dilemma to solve, which will keep occupying scholars' minds. This project attempted to contribute to this debate. It does not pretend to exhaustively explain consumer countries' cooperation in oil. It is merely a stepping stone toward our understanding of these phenomena. Hopefully, future research will bring us even closer.

Notes

1. Stinchcombe, *Constructing Social Theories.*
2. Grieco, *Cooperation Among Nations.*
3. ExxonMobil, "Outlook For Energy: A View to 2030" http://www.exxon mobil.com/Corporate/Files/news_pub_eo_2009.pdf (accessed 15 March 2011); U.S. Energy Information Administration, "International Energy Outlook".
4. ExxonMobil, "Oulook for Energy: A View to 2030."
5. U.S. Energy Information Administration, "World Proved Reserves of Oil, Most Recent Estimates."
6. U.S. Energy Information Administration, "Short-term Energy Outlook: February 10, 2010, Release."
7. Gal Luft, "Fueling the Dragon: China's Race into the Oil Market," Institute for the Analysis of Global Security, http://www.iags.org/china.htm (accessed 12 March 2011).
8. Mills, *The Myth of the Oil Crisis.*
9. Robert L. Larsson, *Sweden and the NEGP: A Pilot Study of the North European Gas Pipeline and Sweden's Dependence on Russian Energy* (Stockholm: Defense Research Agency, 2006).
10. OECD Oil Committee, "The Outlook for Oil Supply and Demand."
11. Lieber, "Europe and America in the World Energy Crisis."
12. U.S. Energy Information Administration, "Short-term Energy Outlook: February 10, 2010, Release."
13. Mills, *The Myth of the Oil Crisis.*
14. Steve Yetiv, *Crude Awakenings: Global Oil Security and American Foreign Policy* (Ithaca, NY: Cornell University Press, 2004).
15. Gal Luft and Anne Korin, "Terrorism Goes to See," *Foreign Affairs* 83, no. 6 (2004).

Bibliography

"Aftermath of a Revolution." *Petroleum Economist*, April 1979, 138–9.

"Arabs to Hike Output to Offset Iran/Iraq Losses." *Oil & Gas Journal*, 13 October 1980, 88.

Atherton, Alfred. "The United States and the Suez Crisis: The Uses and Limits of Diplomacy." In *The Suez-Sinai Crisis 1956: Retrospective and Reappraisal*, edited by Selwyn Troen and Moshe Shemesh, London: Frank Cass & Co. Ltd., 1990.

Axelrod, Robert M. *The Evolution of Cooperation*. New York: Basic Books, 1984.

Badger, Daniel, and Robert Belgrave. *Oil Supply and Price: What Went Right in 1980?* Paris: Atlantic Institute for International Affairs, 1982.

Bajaj, Vikas, and Anthony Ramirez. "Spot Shortages of Gas Reported Around Country." *New York Times*, 5 September 2005.

Baldwin, David A. "Power Analysis and World Politics: New Trends versus Old Tendencies." *World Politics* 31, no. 2 (1979): 161–194.

Bar-On, Mordechai. "David Ben-Gurion and the Sevre Collusion " In *Suez 1956: The Crisis and Its Consequences*, edited by W. M. Roger Louis and Roger Owen, New York: Oxford University Press, 1989.

Barber, William. "The Eisenhower Energy Policy: Reluctant Intervention." In *Energy Policy in Perspective*, edited by Craufurd Goodwin, Washington, DC: Brookings Institution, 1981.

Barrow, Martin. "Price of Petrol Under Pressure as Oil Plunges." *Times*, 19 January 1991.

Barsky, Robert B., and Lutz Kilian. *Oil and the Macroeconomy Since the 1970s*. Cambridge, MA: National Bureau of Economic Research, 2004.

Beavis, Simon, Ben Laurance, and Deborah Wise. "Finance and Economics: War in the Gulf—International Energy Agency Activates Contingency Plans by Releasing Extra 2.5m Barrels Day on to the Market " *The Guardian*, 18 January 1991.

Binder, David. "Kissinger Said to Express Disgust at Allies' Position." *New York Times*, 31 October 1973.

"Bisher Keine OECD-Beschlüsse." *Frankfurter Allgemeine Zeitung*, 21 June 1951.

Blair, William M. "Declaration of an Oil Emergency Is Considered by U.S. and Companies." *New York Times*, 9 June 1967.

Blanchard, Christopher M., and Alfred Prados, "Saudi Arabia Terrorist Financing Issues," Congressional Research Service, http://www.fas.org/sgp/crs/mideast (accessed 16 February 2011)

Bossuat, Gerard. "The Marshal Plan and European Integration: Limits of an Ambition." In *The Marshal Plan Today: Model and Metaphor*, edited by John Agnew and J. Nicholas Entrikin. New York: Routledge, 2004.

Bowie, Robert. "Eisenhower, Dulles and the Suez Crisis." In *Suez 1956: The Crisis and Its Consequences*, edited by W. M. Roger Louis and Roger Owen, Oxford: Clarendon Press, 1989.

"BP Suspends Third Party Crude Oil Deliveries." *Oil & Gas Journal*, 3 September 1979.

Carmical, J. H. "Shortage Made Up in Loss of Iranian Oil." *New York Times*, 26 August 1951, 1.

Carmical, J.H. "Foreign Oil Standby US Is Awaited." *New York Times*, 22 October 1951.

Claude, Inis L. *Swords into Plowshares; The Problems and Progress of International Organization*. New York: Random House, 1971.

"The Community and Events in the Middle East," http://aei.pitt.edu/5582/01/002280_1.pdf (accessed 15 January 2011).

"Control of Crude Oil Supplies." *Petroleum Economist*, September 1979, 350–1.

Deese, David A. "Energy: Economics, Politics, and Security." *International Security* 4, no. 3 (1979): 140–153.

Deese, David A., and Joseph S. Nye. *Energy and security*. Cambridge, MA: Ballinger Pub. Co., 1981.

Ditte, Pascal, and Peter Roell. "Past Oil Price Shocks: Political Background and Economic Impact " *ISPSW Publications* (2006). http://www.isn.ethz.ch/isn/Digital-Library/Publications/Detail/?id=20499 (accessed 23 February, 2011).

Downie, Leonard. "Britain to Reduce Oil Output." *Washington Post*, 4 January 1980.

Drozdiak, William. "International Agency Calls Oil Supplies Adequate." *The Washington Post*, 29 September 1990.

Dupuy, Trevor N. *Elusive Victory: the Arab-Israeli Wars, 1947–1974*. New York: Harper and Row, 1978.

Editorial. "IEA Snubs Bush on Strategic Oil." *Oil & Gas Journal*, 8 October 1990.

________. "La France ne connaît pas de difficultés pour le transport du brut qu'elle importe." *Le Monde*, 16 November 1973.

________. "Le boycottage des pays arabes commence à inquiéter les Etats-Unis." *Le Monde*, 24 October 1973.

Elm, Mostafa. *Oil, Power, and Principle: Iran's Oil Nationalization and Its Aftermath*. 1st ed. Contemporary issues in the Middle East. Syracuse, NY: Syracuse University Press, 1992.

"Energy; Not So Crude." *The Economist*, 8 December 1979.

European Commission, "EU Crude Oil Imports," http://ec.europa.eu/energy/observatory/oil/import_export_en.htm (accessed 10 March 2011).

"The European Council, Paris, 12-13 March 1979," http://aei.pitt.edu/1458/01/Paris_march_1979.pdf (accessed 16 December 2010).

Evans, A.C. "European Community Law and the Problem of Oil Shortages." *The International and Comparative Law Quarterly* 31, no. 1 (1982): 1–16.

ExxonMobil, "Outlook for Energy: A View to 2030," http://www.exxonmobil.com/Corporate/Files/news_pub_eo_2009.pdf (accessed 15 March 2011).

Farnsworth, Clyde H. "Busy Europe and Mideast Ports Cast Doubt on Oil Cut's Effect." *New York Times*, 22 December 1973.

________. "France Is Facing Oil Delivery Cut." *New York Times*, 22 November 1973.

________. "Industrial Nations Put Off Crisis Action." *New York Times*, 26 October 1973.

Feld, Lowell S. "Oil Markets in Crisis: Major Oil Supply Disruptions Since 1973." In *The New Global Oil Market: Understanding Energy Issues in the World Economy*, edited by Siamack Shojai. Westport, CT: Greenwood Publishing Group, 1995.

Fletcher, Sam. "Production, Refining Resumes Slowly Following Katrina." *Oil & Gas Journal*, 12 September 2005.

Fontaine, André. "De Suez à Londre." *Le Monde*, 27 September 1956.

Foreign Relations of the United States, 1951. 1985. Vol. IV.

Frankel, P. H. "Oil Supplies During the Suez Crisis - On Meeting a Political Emergency." *The Journal of Industrial Economics* 6, no. 2 (1958): 85–100.

George, Alexander L. "Case Studies and Theory Development: The Method of Structured, Focused Comparison." In *Diplomacy: New Approaches in Theory, History, and Policy*, edited by Paul Gordon Lauren. New York: Free Press, 1979.

Georges-Picot, Jacques. *The Real Suez Crisis: The End of a Great Nineteenth-Century Work*. New York: Harcourt Brace Jovanovich, 1978.
"German Oil." *Washington Post*, 4 January 1980.
"Germany." *Business Week*, 9 July 1979, 36.
Gilpin, Robert. *War and Change in World Politics*. Cambridge, MA: Cambridge University Press, 1981.
Goldman, Peter. "Carter's New Energy Plan." *Newsweek*, 2 April 1979.
Grieco, Joseph. "Anarchy and the Limits of Cooperation: A Realist Critique of the Newest Liberal Institutionalism." *International Organization* 42, no. 3 (1988): 485–507.
Grieco, Joseph M. *Cooperation Among Nations: Europe, America, and Non-tariff Barriers to Trade*. Ithaca, NY: Cornell University Press, 1990.
Griffiths, Richard. "An Act of Creative Leadership: The End of the OEEC and the Birth of the OECD." In *Explorations in OEEC History*, edited by Richard Griffiths. Paris: OECD Publishing, 1997.
Haas, Ernst B. *The Uniting of Europe; Political, Social, and Economic Forces, 1950-1957* The Library of World Affairs, no. 42. Stanford, CA: Stanford University Press, 1958.
Haggard, Stephan, and Beth A. Simmons. "Theories of International Regimes." *International Organization* 41, no. 3 (1987): 491–517.
Hamilton, James D. "Oil and the Macroeconomy Since World War II." *The Journal of Political Economy* 91, no. 2 (1983): 228–248.
Heisbourg, Francois. "France and the Gulf Crisis." In *Western Europe and the Gulf*, edited by Nicole Gnesotto and John Roper. Paris: Institute for Security Studies of WEU, 1992.
Hellema, Duco, Cees Wiebes, and Gerardus Tobias Witte. *The Netherlands and the Oil Crisis: Business as Usual*. Amsterdam: Amsterdam University Press, 2004.
Hershey, Robert D. "War Effect on Energy Minimized." *New York Times*, 28 December 1990.
Hiro, Dilip. *Desert Shield to Desert Storm: The Second Gulf War*. New York: Authors Choice Press, 2003.
"How Scarce Is Oil?" *The Economist*, 15 December 1973.
Hurrell, Andrew. "Hegemony, Liberalism and Global Order: What Space for Would-be Great Powers?" *International Affairs* 82, no. 1 (2006): 1–19.
Ibrahim, Youssef. "Bush's Move on Stockpile Brings Some Dissent Overseas." *New York Times*, 28 September 1990.
Ikenberry, G. John. "The Irony of State Strength: Comparative Responses to the Oil Shocks in the 1970s." *International Organization* 40, no. 1 (1986): 105–137.
Ikenberry, Gilford John. *America Unrivaled: The Future of the Balance of Power*. Cornell Studies in Security Affairs. Ithaca, NY: Cornell University Press, 2002.
International Energy Agency. "Analysis of the Impact of High Oil Prices on the Global Economy." (2004). http://www.iea.org/papers/2004/high_oil_prices.pdf [accessed 23 February 2011].
________. *Energy Policies and Programs of IEA Countries*. Paris: IEA, 1980. http://www.iea.org/publications/free_new_Desc.asp?PUBS_ID=1271 (accessed 15 March, 2011).
________. *Monthly Oil Market Report: April*. Paris: IEA, 1991.
________. *Monthly Oil Market Report: February*. Paris: IEA, 1991.
________. *Monthly Oil Market Report: January*. Paris: IEA, 1991.
________. *Monthly Oil Market Report:October*. Paris: IEA, 1990.
________. *Oil Information 2002: Donnees Sur Le Petrole*. Paris: IEA, 2002.

"Iraqi Move Adds to OPEC Price Turmoil." *Oil & Gas Journal*, 18 June 1979.

James, Scott C., and David A. Lake. "The Second Face of Hegemony: Britain's Repeal of the Corn Laws and the American Walker Tariff of 1846." *International Organization* 43, no. 01 (1989): 1–29.

Jervis, Robert. "Realism, Neoliberalism, and Cooperation: Understanding the Debate." In *Progress in International Relations Theory: Appraising the Field*, edited by Colin Elman and Miriam Fendius Elman. Cambridge, MA: MIT Press, 2003.

________. "Security Regimes." *International Organization* 36, no. 2 (1982).

Jones, Brendan. "Canal Closing is Assessed." *New York Times*, 16 July 1967.

Kalt, J. P. *Economics and Politics of Oil-price Regulation: Federal Policy in the Post-Embargo Era.* MIT Press Series on the Regulation of Economic Activity, 1981.

Kapstein, Ethan. *The Insecure Alliance: Energy Crises and Western Politics Since 1944.* New York: Oxford University Press, 1990.

Keddie, Nikki R., and Yann Richard. *Modern Iran: Roots and Results of Revolution.* New Haven, CT: Yale University Press, 2003.

Kemezis, Paul. "Dutch Ask Help As Arabs Cut Oil." *New York Times*, 31 October 1973.

Keohane, Robert. "Theory of World Politics: Structural Realism and Beyond." In *Neorealism and Its Critics*, edited by Robert Keohane. New York: Columbia University Press, 1986.

Keohane, Robert, and Lisa Martin. "Institutional Theory as a Research Program." In *Progress in International Relations Theory: Appraising the Field*, edited by Colin Elman and Miriam Fendius Elman. Cambridge, MA: MIT Press, 2003.

Keohane, Robert O. *After Hegemony: Cooperation and Discord in the World Political Economy.* Princeton, NJ: Princeton University Press, 1984.

________. *International Institutions and State Power: Essays in International Relations Theory.* Boulder, CO: Westview Press, 1989.

________. "Reciprocity in International Relations." *International Organization* 40, no. 1 (1986): 1–27.

Keohane, Robert O., and Lisa L. Martin. "The Promise of Institutionalist Theory." *International Security* 20, no. 1 (1995): 39.

Keohane, Robert O., and Joseph S. Nye. *Power and Interdependence.* New York: Longman, 2001.

Keohane, Robert O., Joseph S. Nye, and Stanley Hoffmann. *After the Cold War: International Institutions and State Strategies in Europe, 1989–1991.* Cambridge, MA: Harvard University Press, 1993.

Khadduri, Majid, and Edmund Ghareeb. *War in the Gulf, 1990–91: The Iraq-Kuwait Conflict and Its Implications.* New York: Oxford University Press, 1997.

Kilian, Lutz. "Exogenous Oil Supply Shocks: How Big Are They and How Much Do They Matter for the US Economy?" *CEPR Discussion Papers* (2006). http://bankofcanada.ca/en/conference_papers/commodity_price2006/kilian.pdf (accessed 23 February 2011).

Kindleberger, Charles P. *The World in Depression: 1929–1939.* Berkley, CA: University of California Press, 1973.

Knutsen, Torbjuırn L. *The Rise and Fall of World Orders.* Manchester: Manchester University Press, 1999.

Koven, Ronald. "IEA to Study European Oil Buying by US." *Washington Post*, 8 June 1979.

________. "Schlesinger: U.S. Losing Oil to Europe." *Washington Post*, 23 May 1979.

Kraft, John, and Mark Rodekohr. "Crude Oil Price Controls: Their Purpose and Impact." *Denver University Law Review*, 1979.

Krasner, Stephen D. "Global Communications and National Power: Life on the Pareto Frontier." *World Politics*, 1991, 336–366.

Kunz, Diane. "The Importance of Having Money: The Economic Diplomacy of the Suez Crisis." In *Suez 1956: The Crisis and Its Consequences*, edited by W. M. Roger Louis and Roger Owen, Oxford: Clarendon Press, 1989.

Laqueur, Walter. *Confrontation: The Middle East and World Politics*. London: Abacus, 1974.

________. *The Road to War: The Origin and Aftermath of the Arab-Israeli Conflict, 1967/8*. Harmondsworth: Penguin, 1969.

Larsson, Robert L. *Sweden and the NEGP: A Pilot Study of the North European Gas Pipeline and Sweden's Dependence on Russian Energy*. Stockholm: Defense Research Agency, 2006.

"Leading Article: Volatile Oil." *Lloyd's List*, 18 January 1991.

Lenczowski, George. "The Oil-producing Countries." In *The Oil Crisis*, edited by Raymond Vernon. New York: W.W. Norton&Company Inc., 1976.

Lewis, Paul. "Major Companies Plan to Cut Oil Purchases." *New York Times*, 10 December 1980.

Lieber, Robert J. "Energy, Economics and Security in Alliance Perspective." *International Security* 4, no. 4 (1980): 139–163.

________. "Europe and America in the World Energy Crisis." *International Affairs (Royal Institute of International Affairs 1944-)* 55, no. 4 (1979): 531–545.

________. "Oil and Power after the Gulf War." *International Security* 17, no. 1 (1992): 155–176.

________. *Oil and the Middle East War—Europe in the Energy Crisis*. Harvard Studies in International Affairs, no. 35. Cambridge: Center for International Affairs, Harvard University, 1976.

________. *The Oil Decade: Conflict and Cooperation in the West*. New York: Praeger Publishers, 1983.

Lipson, Charles. "International Cooperation in Economic and Security Affairs." *World Politics*, 1984, 1–23.

________. "The Transformation of Trade: The Sources and Effects of Regime Change." *International Organization* 36, no. 2 (1982): 417–455.

Luft, Gal, "Fueling the Dragon: China's Race Into the Oil Market", Institute For the Analysis of Global Security, http://www.iags.org/china.htm (accessed 12 March 2011).

Luft, Gal, and Anne Korin. "Terrorism Goes to See." *Foreign Affairs* 83, no. 6 (2004): 61–71.

Mansfield, Edward D. "The Concentration of Capabilities and International Trade." *International Organization* 46, no. 3 (1992): 731–764.

Matthews, Ken. *The Gulf Conflict and International Relations*. London; New York: Routledge, 1993.

McGirk, Tim. "How Far Will Iraq Go." *Christian Science Monitor,* 25 September 1980.

Mearsheimer, John J. "The False Promise of International Institutions." *International Security* 19, no. 3 (1994): 5–49.

________. *The Tragedy of Great Power Politics*. New York: W.W. Norton, 2003.

Michaelis, Alfred. "The Middle East Conflict and Oil Supplies." *Frankfurter Allgemeine Zeitung,* 7 November 1956.

Milani, Mohsen M. *The Making of Iran's Islamic Revolution: From Monarchy to Islamic Republic*. Westview Special Studies on the Middle East. Boulder, CO: Westview Press, 1988.

Mills, Robin M. *The Myth of the Oil Crisis: Overcoming the Challenges of Depletion, Geopolitics, and Global Warming*. Westport, CT: Praeger, 2008.
Milner, Helen V. "Power, Interdependence, and Nonstate Actors in World Politics: Research Frontiers." In *Power, Interdependence, and Nonstate Actors in World Politics*, edited by Helen V. Milner and Andrew Moravcsik. Princeton, NJ: Princeton University Press, 2009.
"Minutes of the 3rd Meeting of the OECD Oil Committee." Paris: Organization for Economic Cooperation and Development, 1962.
"Minutes of the 9th Session of the OECD Oil Committee." Paris: Organization for Economic Cooperation and Development, 1966.
"Minutes of the 11th Session of the OECD Oil Committee." Paris: Organization for Economic Cooperation and Development, 1967.
"Minutes of the 26th OECD Oil Committee Meeting." Paris: Organization for Economic Cooperation and Development, 1973.
"Minutes of the 27th OECD Oil Committee Meeting." Paris: Organization for Economic Cooperation and Development, 1973.
"Minutes of the 60th Meeting of the OEEC Oil Committee." Paris: Organization for Europan Economic Cooperation, 1951.
"Minutes of the 77th OEEC Oil Committee Meeting." Organization for European Economic Cooperation, 1956.
"Minutes of the 91st OEEC Oil Committee Meeting." Organization for European Economic Cooperation, 1957.
"Minutes of the 107th, 108th, 109th and 110th OEEC Oil Committee Meetings." Paris: Organization for European Economic Cooperation, 1960.
"Minutes of the 116th OEEC Oil Committee Meeting." Paris: Organization for European Economic Cooperation, 1961.
Mohr, Charles. "Carter Asks Strict Fuel Savings." *New York Times*, 19 April 1977.
Moravcsik, Andrew. "Taking Preferences Seriously: A Liberal Theory of International Politics." *International Organization* 51, no. 4 (1997): 513–553.
Nash, N. "Pressure Is Mounting to Tap U.S. Oil Stockpile." *New Yrok Times*, 18 September 1990.
"News in Brief: United Kingdom." *Petroleum Economist*, October 1979, 435.
Nossiter, Bernard D. "U.S. and Allies Discuss Joint Fleet to Protect Vital Strait of Hormuz." *New York Times*, 25 September 1980.
Nuttall, Simon J. *European Political Co-operation*. Oxford: Oxford University Press, 1992.
Nye, Joseph. "Energy Nightmares." *Foreign Policy*, 1980, 132-154.
Odell, Peter R. *Oil and World Power*. Harmondsworth, Middlesex: Penguin Books, 1986.
OECD Oil Committee. "The Outlook for Oil Supply and Demand." Paris: Organization for Economic Cooperation and Development, 1967.
Oil, High Level Group of the Special Committee for. "Draft Preliminary Report on the 1967 Oil Emergency." Paris: OECD, 1968.
Organisation for Economic Cooperation and Development. *Oil; The Present Situation and Future Prospects*. Paris: OECD, 1973.
Organisation for European Economic Cooperation. *Europe's Growing Needs of Energy: How Can They Be Met?* Paris: OEEC, 1956.
Organization for European Economic Cooperation. *Report by the Oil Committee to Be Published in 1956*. Paris: OEEC, 1956.
Oye, Kenneth A. *Cooperation Under Anarchy*. Princeton, NJ: Princeton University Press, 1986.

Palango, Paulo. "Canada Must Suffer Shortages Like Other Nations, Oil Man Says." *The Globe and Mail*, 16 February 1979.

Parisi, Anthony J. "Attack on Oil Sharing Stirs Concern in Industry." *New York Times*, 17 February 1979.

Pattner, Steven. "A New Distribution in World Oil Sales Perils U.S. Imports." *The New York Times*, 15 March 1979.

Paully, David. "Over the Oil Barrel." *Newsweek*, 9 July 1979.

Percy, Charles. "To Meet an Oil Crisis." *New York Times*, 30 September 1980.

Phillips, James A. *The Iranian Oil Crisis*. Washington, D.C.: Heritage Foundation, 1979.

Potts, Mark, and Thomas W. Lippman. "Oil Prices Tumble on World Markets." *Washington Post*, 17 January 1991.

Press, Associated. "Europe not Alarmed by Loss of Iranian Oil." *New York Times*, 24 August 1951.

________. "Joint Oil Compact Authorized by US." *New York Times*, 3 August 1951.

Prodi, Romano, and Alberto Clo. "Europe." In *The Oil Crisis*, edited by Raymond Vernon. New York: W.W. Norton&Company, 1976.

Randall, Stephen J. *United States Foreign Oil Policy since World War I: For Profits and Security*. Montreal: McGill-Queen's University Press, 2005.

Richardson, Louise. *When Allies Differ: Anglo-American Relations During the Suez and Falklands Crises*. 1st ed. New York: St. Martin's Press, 1996.

"Roy Jenkins, President of the Commission of the European Communities: Address to the European Parliament, 19 July 1979," http://aei.pitt.edu/11341/01/48-Jenkins-07-19-1979.pdf (accessed 13 January 2011).

Rubin, Barry M. *Paved with Good Intentions: The American Experience and Iran*. New York: Oxford University Press, 1980.

Ruggie, John Gerard. "What Makes the World Hang Together? Neo-Utilitarianism and the Social Constructivist Challenge." *International Organization* 52, no. 4 (1998): 855–885.

Safran, Nadav. *From War to War:The Arab-Israeli Confrontation, 1948–1967: A Study of the Conflict from the Perspective of Coercion in the Context of Inter-Arab and Big Power Relations*. Indianapolis: Pegasus, 1969.

"Saudis List Customers for Extra Crude." *Oil & Gas Journal*, 3 November 1980.

Schneider, Martin. "The Impact of Oil Price Changes on Growth and Inflation." *Monetary Policy and the Economy*, no. 2 (2004). http://www.oenb.at/en/img/mop_20042_the_impact_tcm16-19678.pdf (accessed 23 February, 2011).

Scott, Richard. *The History of the International Energy Agency, 1974–1994: IEA, the First 20 Years*. Paris: OECD Publications and Information Centre, 1994.

Shuster, Alvin. "Alert Puzzles Europeans." *New York Times*, 27 October 1973.

Staff Report to the Committee on Foreign Relations. *War in the Persian Gulf: The U.S. Takes Sides*. Washington, D.C.: U.S. Senate, 1987.

Sterner, Michael. "The Iran-Iraq War." *Foreign Affairs* 63, no. 1 (1984): 128–143.

Stinchcombe, Arthur L. *Constructing Social Theories*. New York: Harcourt, Brace & World, 1968.

Stobaugh, Robert B. "The Oil Companies in the Crisis." In *The Oil Crisis*, edited by Raymond Vernon. New York: W. W. Norton&Company Inc, 1976.

Strange, Susan. *The Retreat of the State: The Diffusion of Power in the World Economy*, Cambridge studies in international relations, 49. New York: Cambridge University Press, 1996.

Tagliabue, John. "Spot Oil Price Climbing as War Pares Supplies." *New York Times*, 24 October 1980.

Taylor, Peter J. *The Way the Modern World Works: World Hegemony to World Impasse*. Chichester: Wiley, 1996.
Tertzakian, Peter. *A Thousand Barrels a Second: The Coming Oil Break Point and the Challenges Facing an Energy Dependent World*. New York: McGraw-Hill, 2006.
"Third Time Lucky?". *The Economist*, 6 October 1990.
Tibi, Bassam. *Conflict and War in the Middle East, 1967–91: Regional Dynamic and the Superpowers*. New York: St. Martin's Press, 1993.
"Transcript of Carter's Statement On Iran-Iraq War." *New York Times*, 25 September 1980.
Turner, Louis. "The Oil Majors in World Politics." *International Affairs (Royal Institute of International Affairs)* 52, no. 3 (1976): 368–380.
U.S. Congress. House. *Energy Impact of the Persian Gulf Crisis: Hearing Before the Committee on Energy and Commerce*, 101 Cong., 2 sess., 1990.
U.S. Congress. Senate. *Emergency Oil Lift Program and Related Oil Problems: Joint Hearings Before Subcommittees of the Committee on the Judiciary and Committee on Interior and Insular Affairs*, 85 Cong.,1 sess., 1957.
________. *Multinational Corporations and U.S. Foreign Policy. Part 4: Multinational Petroleum Companies and Foreign Policy: Hearings Before the Subcommittee on Multinational Corporations, Committee on Foreign Relations*, 93 Cong., 2 sess., 1974.
U.S. Energy Information Administration, "Energy Perspectives," http://www.eia.doe.gov/emeu/aer/txt/ptb1105.html (accessed 15 November 2010).
________, "International Energy Outlook" http://www.eia.doe.gov/oiaf/ieo/index.html (accessed 15 February 2011).
________, "International Energy Outlook," http://www.eia.doe.gov/oiaf/archive/ieo05/index.html (accessed 10 March 2011).
________, "International Petroleum Stocks," http://www.eia.doe.gov/emeu/international/oilstocks.html (accessed 30 September 2011).
________, "Major Disruptions of World Oil Supply," http://www.eia.doe.gov/emeu/25opec/sld001.htm (accessed 10 July 2010).
________, "OECD Countries and World Petroleum (Oil) Demand, 1970–2008," http://www.eia.doe.gov/emeu/ipsr/t46.xls (accessed 15 March 2011).
________, "Petroleum Imports by Country of Origin," http://www.eia.doe.gov/emeu/aer/txt/ptb0504.html (accessed 17 December 2010).
________, "Petroleum Overview, 1949–2008," http://www.eia.doe.gov/emeu/aer/txt/ptb05 01.html (accessed 18 November 2010).
________, "Refinery Utilization and Capacity," http://tonto.eia.doe.gov/dnav/pet/pet_pnp_unc_dcu_nus_m.htm (accessed 10 March 2011).
________, "Russia: Oil Exports," http://www.eia.doe.gov/emeu/cabs/Russia/Oil_exports.html (accessed 10 March 2011).
________, "Short-term Energy Outlook: February 10, 2010 Release, " http://www.eia.doe.gov/pub/forecasting/steo/oldsteos/feb10.pdf (accessed 15 February 2010).
________, "This Week in Petroleum," http://tonto.eia.doe.gov/oog/info/twip/twiparch/twiparch.html#2004 (accessed 10 March 2011).
________, "U.S. Field Production of Crude Oil" http://tonto.eia.doe.gov/dnav/pet/hist/LeafHandler.ashx?n=PET&s=MCRFPUS1&f=M (accessed 09 March 2011).
________, "U.S. Product Supplied of Crude Oil and Petroleum Products," http://tonto.eia.doe.gov/dnav/pet/hist/LeafHandler.ashx?n=PET&s=MTTUPUS2&f=M (accessed 13 February 2011).
________, "World Crude Oil Production: OPEC Members," http://tonto.eia.doe.gov/merquery/mer_data.asp?table=T11.01a (accessed 15 March 2011).

________,"World Oil Transit Chokepoints: Suez/Sumed," http://www.eia.doe.gov/cabs/World_Oil_Transit_Chokepoints/Suez.html (accessed 20 February 2011).

________, "World Petroleum Consumption, 1960–2007," http://www.eia.doe.gov/emeu/aer/txt/ptb1110.html (accessed 17 December 2010).

________, "World Proved Reserves of Oil, Most Recent Estimates," http://www.eia.doe.gov/emeu/international/reserves.html (accessed 15 February 2011).

United Press International. "Bedenken Gegen Ölembargo." *Frankfurter Allgemeine Zeitung*, 1 July 1967.

________. "Seasonal Shortage of Heating Oil Seen." *New York Times*, 23 November 1972.

Vaisse, Maurice. "France and the Suez Crisis." In *Suez 1956: The Crisis and Its Consequences*, edited by W. M. Roger Louis and Roger Owen. Oxford: Clarendon Press, 1989.

van Evera, Stephen. "Why Cooperation Failed in 1914." *World Politics* 38, no. 1 (1985): 80–117.

Vincent, L. A., and G. Matthys. "La Situation Economique." *Revue Economique* 2, no. 6 (1951): 777–787.

Wall, Irwin M. *France, the United States, and the Algerian War*. Berkeley: University of California Press, 2001.

Waltz, Kenneth N. "Structural Realism after the Cold War." *International Security* 25, no. 1 (2000): 5–41.

Waltz, Kenneth Neal. *Theory of International Politics*. Reading, MA: Addison-Wesley Pub. Co., 1979.

"War Curbs Oil Exports by Iran, Iraq." *Oil & Gas Journal*, 29 September 1980.

"War in the Oil Fields of Iran and Iraq ". *Newsweek*, 6 October 1980.

Ward, Matthew L. "Price of Oil Dips Below $20 a Barrel." *New York Times*, 19 January 1991.

"Warnings on Oil Shortage at IEA Meeting." *Petroleum Economist*, June 1979, 239.

Watt, Donald Cameron. *Documents on the Suez Crisis, 26 July to 6 November 1956*. London: Royal Institute of International Affairs, 1957.

"What Oil Crisis?" *The Economist*, 11 August 1979.

Wilhelmshaven, F. A. Z. "Erstes Rohöl aus Amerika." *Frankfurter Allgemeine Zeitung*, 25 July 1967.

Willenborg, Robbert, Christophe Tonjes, and Wilbur Perlot. "Europe's Oil Defences: An Analysis of Europe's Oil Supply Vulnerability and Its Emergency Oil Stockholding Systems."(2004), http://www.clingendael.nl/publications/2004/20040100_ciep_paper_willenborg.pdf (accessed 24 February, 2011).

Yergin, Daniel. "Ensuring Energy Security." *Foreign Affairs* 85, no. 2 (2006): 69.

________. *The Prize: The Epic Quest for Oil, Money, and Power*. New York: Simon & Schuster, 1991.

Yetiv, Steve. *Crude Awakenings: Global Oil Security and American Foreign Policy*. Ithaca, NY: Cornell University Press, 2004.

________. *The Persian Gulf Crisis*. Westport, CT: Greenwood Press, 1997.

Index

About the Author

Vessela Chakarova is a lecturer in international relations at Amsterdam University College and a member of the adjunct faculty at Webster University—Leiden. She has also taught at Old Dominion University. She has worked for the European Commission and a number of NGOs.

Her previous research has focused on the role of international organizations in mediation and conflict resolution, the evolution of NATO, and various works on the EU and European politics.

www.ingramcontent.com/pod-product-compliance
Ingram Content Group UK Ltd.
Pitfield, Milton Keynes, MK11 3LW, UK
UKHW041141300926
15471UKWH00011B/92

9 781498 51543